D0138373

Chemical Activities

Teacher Edition

Christie L. Borgford
Oregon Episcopal School

Lee R. Summerlin
University of Alabama at Birmingham

AMERICAN CHEMICAL SOCIETY

WASHINGTON, DC 1988

Library of Congress Cataloging-in-Publication Data

Borgford, Christie L., 1943-
Chemical activities: Teacher edition/Christie L. Borgford, Lee R. Summerlin.

p. cm.
Bibliography: p.
Includes indexes.

ISBN 0-8412-1416-6 (Teacher ed.)
ISBN 0-8412-1417-4 (Student ed.)

1. Chemistry—Experiments. I. Summerlin, Lee R. II. Title.

QD43.B63 1988
540'.78—dc19 87-28992
CIP

Karen L. McCeney: Copy editing
Paula M. Bérard: Production and indexing
Janet S. Dodd: Managing editor

Carla L. Clemens: Cover design and section opening pages
Dana Borgford: Illustrations in text

Typesetting: Hot Type Ltd., Washington, DC
Typeface: Melior
Printing: Mack Printing Company, Easton, PA
Binding: Nicholstone Book Bindery, Nashville, TN

About the Authors

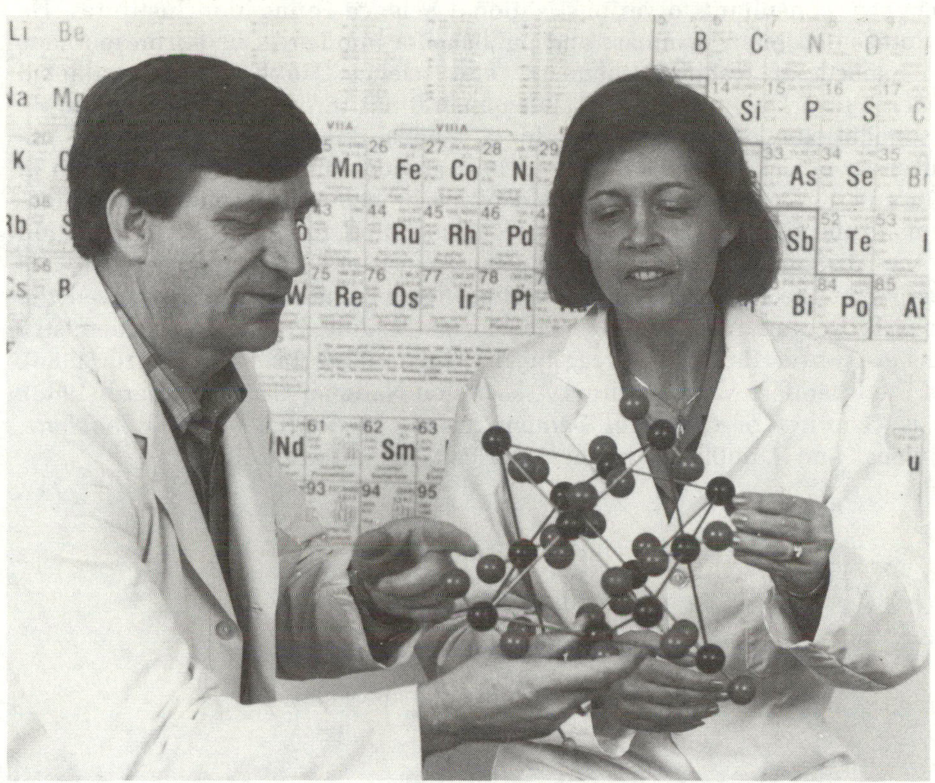

CHRISTIE L. BORGFORD (right) received her B.S. from the University of Washington and her master's degree from the University of Alabama at Birmingham. She taught chemistry in Washington, MO, and is now chair of the Science Department at Oregon Episcopal School in Portland, OR. Throughout her teaching career, she has emphasized the experimental approach to chemistry and has developed and used demonstrations and activities for junior high school students, chemistry students, and chemistry teacher education programs around the country. In addition to the many workshops and courses that she has organized and conducted for science teachers in the Northwest, she was codirector of the Chemical Demonstrations Program for the Institute for Chemical Education (I.C.E.) program at the University of California—Berkeley in 1985 and was director of the laboratory program in 1986 and 1987. She serves as a consultant to several firms and organizations and has published articles and books on chemistry and chemical education, including a laboratory manual for the Berkeley I.C.E. program. She is coauthor of *Chemical Demonstrations: A Sourcebook for Teachers, Volume II,* published by the American Chemical Society. She is active in the American Chemical Society, where she chairs the Division of Chemical Education Prehigh School Science Task Force and is a member of other national ACS committees. She serves her local section as Precollege Chair.

LEE R. SUMMERLIN (left) received his B.A. from Samford University, his M.S. from Birmingham-Southern College, and his Ph.D. from the University of Maryland. He has held many teaching and administrative positions and has served as a consultant to various organizations, companies, colleges, and school boards in this country and abroad. He has served as a peer review panelist for several science education development programs and as a chemistry consultant to various National Science Foundation Institutes. He has presented many seminars and published a number of books on methods and aspects of teaching chemistry and science. He has also conducted numerous workshops on chemical demonstrations throughout the country. He was codirector of the Institute for Chemical Education program at the University of California—Berkeley in 1985 and coordinator for the program in 1986 and 1987. Summerlin has held office or had major committee assignments in the National Science Teachers Association, the American Association for the Advancement of Science, and the American Chemical Society. He is a member of the ACS Task Force on High School Chemistry. He received the James B. Conant Award (1969), the Florida Section Outstanding Chemistry Teacher Award (1967), the Gregg Ingalls Outstanding Teaching Award (1985), and the Chemical Manufacturers Association National Catalyst Award (1986). He is coauthor of *Chemical Demonstrations: A Sourcebook for Teachers,* Volumes I and II, published by the American Chemical Society.

Our book is dedicated

to Marjorie Gardner,

Director of the Lawrence Hall of Science,

University of California—Berkeley.

Her support and encouragement

made this volume possible.

Contents

Chemical Energy and Rates of Reaction

Chemistry Around the House

Chemistry and the Environment

Biochemistry: Chemistry of Living Things

Chemistry of Foods

Chemical Detectives: Tools and Techniques of the Chemist

Kitchen Chemistry

Appendixes

Preface

This book has been prepared in response to the many requests from science teachers for chemical activities that are motivating, simple, inexpensive, appropriate, relevant, safe, and fun. We hope you will find that the activities in this book are all these things and that you and your students will enjoy doing chemistry.

The activities also reflect our philosophy regarding the teaching of science. Students at the junior high school and high school levels should appreciate the historical aspects of chemistry. An understanding of chemistry contributes to their roles as knowledgeable consumers and informed citizens. They should appreciate the role of chemistry in solving environmental and social problems. We also hope that these activities will be extended into home experiments, demonstrations, small group projects, and science fair and science club activities. Students need to understand the experimental nature of chemistry. Predicting, collecting, and organizing data, identifying and controlling variables, and searching for regularities and relationships are all skills required in chemistry. These, along with the manipulative skills of filtering, titrating, and distilling, give the students the tools they need to understand chemistry. Most importantly, we want students to have successful experiences and to enjoy chemistry.

This book has a unique design. The activities are written for use by the student. You may use them as presented or modify them to fit a particular teaching format or situation. The activities are designed to introduce students to chemistry; they are not intended to give complete coverage of each chemical topic in each activity. Encourage students to seek additional information and conduct additional experiments in class. For each activity, we include a background section for the teacher. This section gives some additional information about the activity, some helpful teaching tips, and complete instructions for preparing and using all solutions. Safety notes and special precautions are included here as well.

Most of the activities can be used to introduce or reinforce several chemical topics. Our arrangement of activities in the table of contents represents only one format. A cross reference of activities and chemical topics is given in Appendix 1A. We suggest that you use this listing to find activities for the chemical topic you wish to cover. Also included (Appendix 1B) is a cross reference of laboratory process skills to help you find activities that are specifically designed for skill development. Finally, activities are also cross referenced (Appendix 1C) with the table of contents of *CHEMCOM* (*Chemistry in the Community*), the curriculum material developed with the support of the American Chemical Society.

Among the goals of a good laboratory program is the goal of helping students to learn safe and healthful handling of substances, including household chemicals, gases, and acids and bases. Because the safety and health of students are our first priorities, we have included chemical activities that present a minimum of risk. We have focused on cautious practices throughout the book and include both a "Be Careful" symbol where it is warranted in the procedure and a safety note in the Teaching Tips section. We call your attention to the laboratory safety guidelines, "Always Be Careful in the Laboratory" on the inside front cover and to Appendix 4A, "Safe Use of Chemicals" and Appendix 4B, "Chemical Disposal and Spill Guidelines". These sections describe good laboratory safety procedures and proper use and disposal of the substances used in activities in this book.

Share with students the relevant information found on a Material Safety Data Sheet (MSDS) to raise their awareness of accepted practices for use and disposal of chemicals (*see* Appendix 4A). Make them available to interested students in a designated area, not locked up.

Appendix 5 contains useful resources for teaching science. It includes two safety references, *Safety in Academic Chemistry Laboratories* and "School Science Laboratories, a Guide to Some Hazardous Substances".

Acknowledgments

We appreciate the efforts of the ACS Ad Hoc Committee on Safety: Gary Long, the chair of the committee; James A. Kaufman; W. H. Norton; Douglas Walters; and Jay Young. We thank the ACS Books Department staff, especially Carla L. Clemens, Karen L. McCeney, Paula M. Bérard, and Janet S. Dodd for their help in producing this book in a timely fashion.

We are indebted to many people who have used these activities, offered suggestions, and made corrections. We especially thank J. A. Campbell, recently retired from Harvey Mudd College, for his careful and thorough review of the manuscript and the many suggestions he offered for improvements. We appreciate Joe Burns, assistant professor of science education at the University of Alabama at Birmingham, and Bruce Brown, professor of chemistry at Portland State University, for their helpful suggestions. We also thank the many teachers in our chemistry programs at the University of Alabama at Birmingham, Portland State University, and the University of California—Berkeley, who have tried these activities over the past several years. We are also indebted to the secretarial staff at the University of Alabama at Birmingham who prepared the manuscript: Niki Bettinger, Jackie Holder, Candi Smith, Vicki Orr, and Jean Holt. We thank Dana Borgford for the art within the text. We extend special thanks to the wonderful students at Oregon Episcopal School who shared with us the enthusiasm and excitement of this project.

We hope you and your students will enjoy doing these activities in the spirit of investigation and discovery.

Christie L. Borgford
Oregon Episcopal School
Portland, OR 97223

Lee R. Summerlin
University of Alabama at Birmingham
Birmingham, AL 35294

Preface to the Student Edition

We would like to introduce you to this book by offering a challenge: While you are reading this, think of something around you that does *not* involve chemistry. What did you think of? The air? Air consists of many chemical elements that are undergoing many chemical reactions. In fact, the reaction of ozone in the stratosphere with damaging ultraviolet rays from the sun allows life to exist on the earth. How about your clothing? It is either natural chemical compounds (such as wool, cotton, or linen) or synthetic compounds made in the laboratory. Did you think of yourself? The human body is really nothing but a bag of thousands of chemicals and chemical reactions. Our growth, our senses, even our intelligence is due to chemistry. Whatever you thought of involves chemistry because *everything* that exists involves chemistry.

Chemistry is the study of matter. Scientists today know a great deal about matter because our search for an understanding of things around us has led over the centuries to the accumulation of many observations on the behavior of matter. Observations are extremely important to the chemist. They are so important that chemistry—like other sciences—is generally conducted in a situation where conditions can be controlled and observations can be carefully made and recorded. This place is the chemical laboratory. All the information that we have about chemistry, all the facts and principles that you will learn in your science classes, were the results from experiments made in laboratories.

We want you to share in the excitement of laboratory chemistry. The activities in this book are designed to introduce you to some of the fascinating and exciting chemical reactions that are important to your everyday life. They are intended to spark your curiosity and make you want to learn more about chemistry. They are intended to allow you to work and think like a chemist.

If you are to do this effectively, you must assume some responsibilities, the same responsibilities that professional scientists assume. You must realize that the laboratory is a special place for serious work. You must follow directions and always work in a neat and orderly surrounding. You must realize that chemicals should be used properly and should be handled carefully even though most of the chemicals used in these activities are familiar to you and are found in your home. You must always consider your safety and the safety of those around you as an important part of any laboratory activity. We have given you some special precautions to follow while working in the laboratory. These are found on the inside front cover of the book. You should become familiar with these rules and always follow them when you are working in the laboratory. Finally, you should always remember that the laboratory is a place where observations are made and explanations for these observations sought.

Among the goals of a good laboratory program is learning safe and healthful handling of substances, including household chemicals, gases, and acids and bases. We have focused on cautious practices throughout the book and include a "Be Careful" symbol where it is warranted in the procedure. We call your attention to Appendix 4A, "Safe Use of Chemicals" and Appendix 4B, "Chemical Disposal and Spill Guidelines". These sections describe good laboratory safety procedures and proper use and disposal of the substances used in activities in this book.

Your teacher will tell you about the information found on a Material Safety Data Sheet (MSDS). The information tells about accepted practices for use and disposal of chemicals (*see* Appendix 4A). Your teacher will make them available to you in a designated area.

We especially want you to enjoy chemistry. The activities in this book are designed to be enjoyable and to provide a thrill of discovery. They are designed to give you a better understanding of the things around you and pose some problems that no one has yet answered. Perhaps you will provide some of these answers.

Acknowledgments

We appreciate the efforts of the ACS Ad Hoc Committee on Safety: Gary Long, the chair of the committee, James A. Kaufman, W. H. Norton, Douglas Walters, and Jay Young. We thank the ACS Books Department staff, especially Carla L. Clemens, Karen L. McCeney, Paula M. Bérard, and Janet S. Dodd for their help in producing this book in a timely fashion.

We are indebted to many people who have used these activities, offered suggestions, and made corrections. We especially thank J. A. Campbell, recently retired from Harvey Mudd College, for his careful and thorough review of the manuscript and the many suggestions he offered for improvements. We appreciate Joe Burns, assistant professor of science education at the University of Alabama at Birmingham, and Bruce Brown, professor of chemistry at Portland State University, for their helpful suggestions. We also thank the many teachers in our chemistry programs at the University of Alabama at Birmingham, Portland State University, and the University of California—Berkeley, who have tried these activities over the past several years. We are also indebted to the secretarial staff at the University of Alabama at Birmingham, who prepared the manuscript: Niki Bettinger, Jackie Holder, Candi Smith, Vicki Orr, and Jean Holt. We thank Dana Borgford for all of the art within the text. We extend special thanks to all of those wonderful students at Oregon Episcopal School who shared with us the enthusiasm and excitement of this project.

Christie L. Borgford
Oregon Episcopal School
Portland, OR 97223

Lee R. Summerlin
University of Alabama at Birmingham
Birmingham, AL 35294

Chemistry of Matter: Gases

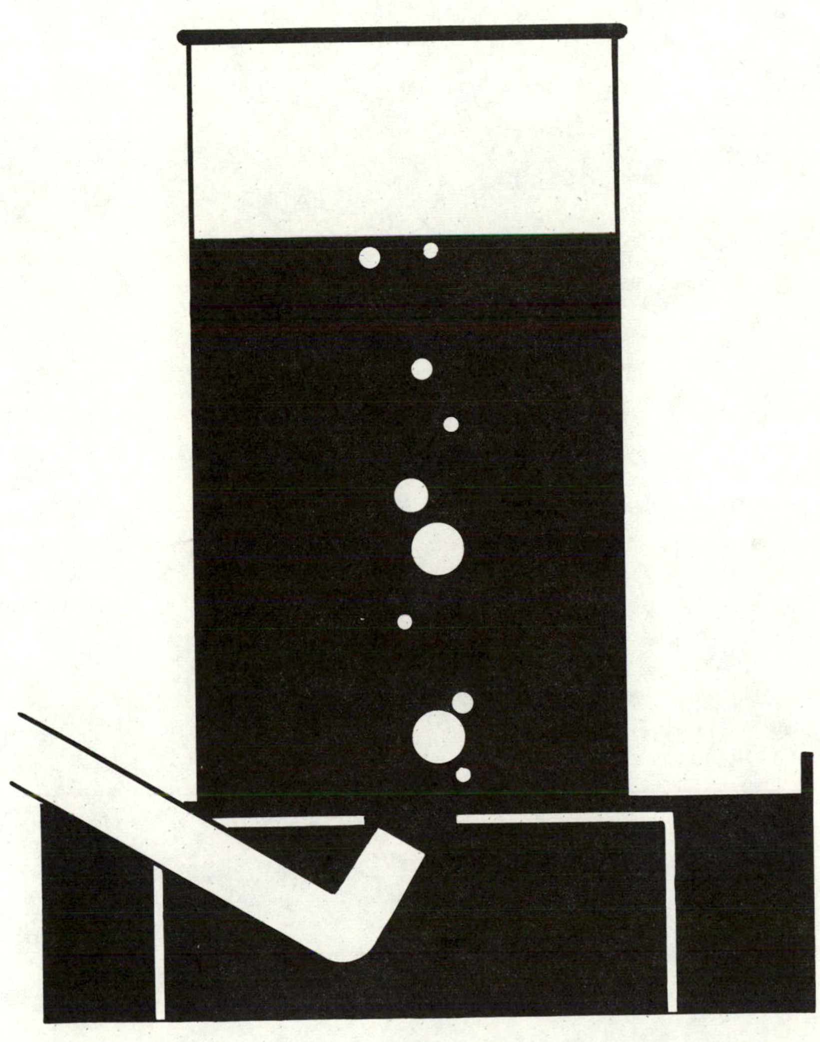

1. Pressure of Air Around Us

You know that air is all around us, and you may know that air consists of molecules of gases that are in constant motion. The constant motion of these particles produces a pressure that we call *atmospheric pressure*. Fortunately, air pressure is exerted in all directions. If air pressure were exerted only from above, for example, we would be instantly crushed by tons of air! In this activity, you will demonstrate the effect of atmospheric pressure.

Materials

1. Small board.
2. Newspaper.
3. Cup.
4. Index card or small piece of cardboard.

Procedure

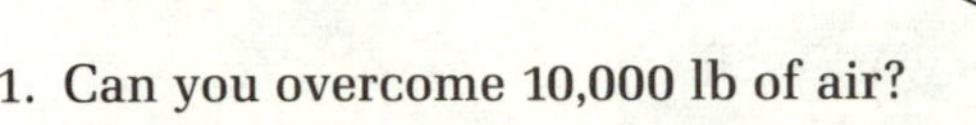

1. Can you overcome 10,000 lb of air?
   A. Open a full section of a newspaper and place it face down on a table or desk top.
   B. Slide a small board beneath the newspaper, but let about one-third of the board hang over the edge of the table or desk.
   C. Slowly push down on the board. Notice that the newspapers will be lifted. Air is able to get under the paper and help lift the paper.
   D. Bring the board back to its position with part overhanging the table and the other end completely covered with the opened newspaper.
   E. Be sure that you have on your safety goggles and that no one is near the table. In one quick movement, strike the overhanging end of the board with your fist. Pretend that your fist is a hammer. Describe the results. Was the newspaper lifted from the table? Why? What happened to the board? Why?
2. Air pressure is exerted in all directions.
   A. Fill a glass completely full of water.
   B. Cover the glass with an index card or a piece of thin cardboard.
   C. Hold the card in place with two fingers, and quickly turn the glass over.
   D. Remove your fingers from the card. What do you observe? Why?

Reactions

1. Atmospheric air exerts a pressure of approximately 15 lb/in.^{2}. Therefore, when spread open, the newspaper has almost 10,000 lb of air pressing on it! If you slowly lift the paper, allowing time for air to flow beneath it, the pressure will be equal on both sides and will not hold the paper in place. However, if you suddenly strike the board beneath the paper in an attempt to raise the paper, the tremendous pressure of atmospheric air on the upper surface will hold the paper in place. In fact, the board may break, but you could never counteract the great pressure holding the paper in place.
2. The card will stay in place, and the water in the glass will not spill. The mass of water in the glass is perhaps 300–400 g (less than 1 lb), depending upon the volume of the glass. This mass is a considerable amount, but the mass pushing up on the cardboard is equivalent to approximately 200–300 lb! If the card falls off, it is probably because the rim of the glass is not even and the seal between the glass and card is not tight.

Questions

1. Why aren't you crushed by the many tons of air pushing down on you?
2. What is air pressure?
3. In Procedure 2, will the card fall off if it is held sideways instead of upside down? (Try it!)

Notes for the Teacher

BACKGROUND

Much of chemistry, especially the work with gases, requires an understanding of the atmosphere and the pressure that it exerts on the earth's surface. These activities are simple but offer dramatic evidence of the tremendous pressure of air.

TEACHING TIPS

1. In Procedure 1, you will find that slats from wooden boxes work well.
2. Students must hit the board with a sharp blow rather than push on it.
3. Using the general rule that the atmosphere exerts about 15 lb of pressure per square inch, have students calculate the force on various objects, for example, the tops of their desks, their books, and the room. Note that pressure units, pounds per square inch, include the unit area. Thus, we can speak of a pressure of 15 lb/in.^{2} or of a total force; for example, a newspaper experiences 10,000 lb of force if it has an area of 672 in.^{2}.
4. We observe the effect of air pressure when liquid travels up a straw because sucking on the straw reduces the air pressure in the straw. Because atmospheric pressure is greater than this reduced pressure, it forces liquid to rise in the straw.
5. Be prepared for the board to break even with the table if it is struck hard enough.
6. If the card falls from the glass tumbler, it is probably because the rim is not even. Try using a small amount of petroleum jelly on the rim.

ANSWERS TO THE QUESTIONS

1. The air pressure inside the body is the same as that outside and is exerted in all directions. Thus, an equal push on all parts of the body occurs and makes us unaware of the pressure.
2. The pressure exerted by the atmosphere. It is measured to be about 15 lb/in.^{2}.
3. No, the card should not fall off if the glass is held sideways.

2. How Much Gas Is in a Bottle of Cola?

We know that bottled colas contain gas because they often foam when they are warm. In fact, the dissolved gas gives colas some of their distinctive and refreshing taste. In this activity we will estimate how much gas a bottle of soft drink contains. We will also test to show that the gas is carbon dioxide.

Materials

1. Ice-cold bottles (12 oz) of various soft drinks.
2. One-hole stopper to fit the soft drink bottle.
3. Small glass tubing to fit the stopper hole.
4. Piece of rubber tubing to fit the glass tubing in the stopper (about 16 in. long).
5. Large plastic tub or basin.
6. Any large clear container or bottle (at least 1 L) with a narrow neck.
7. Large beaker of hot water.
8. Limewater [saturated calcium hydroxide ($Ca(OH)_2$) solution].

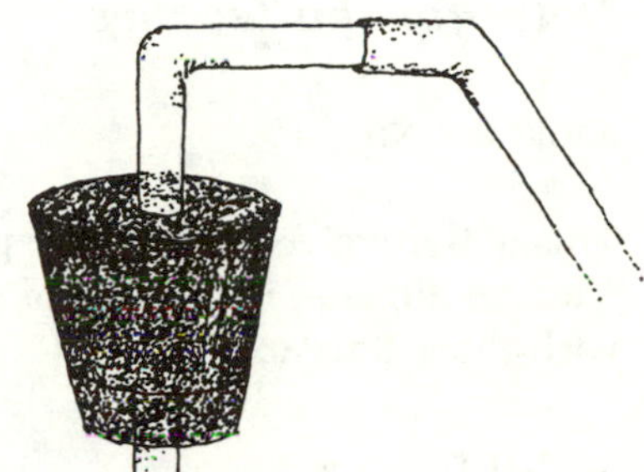

Procedure

1. Obtain a stopper assembly for collecting gas.
2. Fill a large plastic tub one third full of water.
3. Obtain an ice-cold bottle of soft drink.
4. Fill a large container with water and carefully invert the full container in a large plastic tub of water. The container will be heavy, and someone must hold it inverted.
5. Remove the cap from the ice-cold bottle of soft drink. Immediately insert the stopper and place the other end of the rubber tubing beneath and inside the open end of the inverted bottle. The tubing should reach all the way into the bottle.
6. Place the soft drink bottle inside a large beaker of hot (but not boiling) water.
7. Observe the production and collection of gas from the soft drink. Shake the soft drink occasionally to dislodge air bubbles and speed the process.
8. When the production of gas has stopped, mark the level of water in the inverted container with a wax pencil or tape. However, do not remove the container until you are ready for the next step.
9. Obtain about 50 mL of clear limewater. Remove the container, turn it right side up, and pour in the limewater. Cover the mouth of the container with your hand and shake it several times to mix the gas with the limewater. What do you observe? Empty the container.
10. Take another 25-mL sample of clear limewater. Blow your breath, which contains carbon dioxide, into the limewater as you swirl the sample to mix the gas (breath) and liquid. What do you observe? Can you identify the gas that came from the soft drink?
11. Add water to the container until it reaches the wax mark. Measure this volume of water. Is this also the volume of gas? How much gas did you collect from the soft drink?

Reaction

Carbon dioxide is more soluble in cold water than in warm water, as are all gases. Thus, when the soft drink is heated, the carbon dioxide becomes less soluble and leaves the liquid.

Carbon dioxide from the breath and from the soft drink bottle reacts with limewater (calcium hydroxide solution) to form the white solid, calcium carbonate. This reaction is the characteristic test for carbon dioxide:

$$Ca(OH)_2(aq) + CO_2(g) \longrightarrow CaCO_3(s) + H_2O(\ell)$$

Questions

1. How much gas did you collect from your bottle of soft drink? How does this amount compare with the amount collected by other students?
2. What is the gas? How did you prove your conclusion?
3. What can you say about the solubility of a gas as temperature changes?

Notes for the Teacher

BACKGROUND

This activity gives students experience with collecting and identifying the gas (carbon dioxide) in a bottle of a carbonated beverage. They can compare this gas with the gas we exhale.

SOLUTIONS

Prepare the limewater, calcium hydroxide solution, by adding 1 heaping tablespoon of calcium oxide (or calcium hydroxide) to about 1 L of water. Stir the solution, and let the undissolved solid settle to the bottom and pour off the clear solution on top.

$$CaO(s) + H_2O(\ell) \longrightarrow Ca(OH)_2(aq)$$

TEACHING TIPS

SAFETY NOTE: It is best to prepare the stopper and tubing assembly for the students. Fire polish each end of the glass tubing. (Heat it carefully in a hot flame until the sharp ends are smooth.) Lubricate the tubing with glycerol or water and CAREFULLY insert the tubing in the stopper with a twisting motion. You may make a bend before inserting the glass in the stopper.

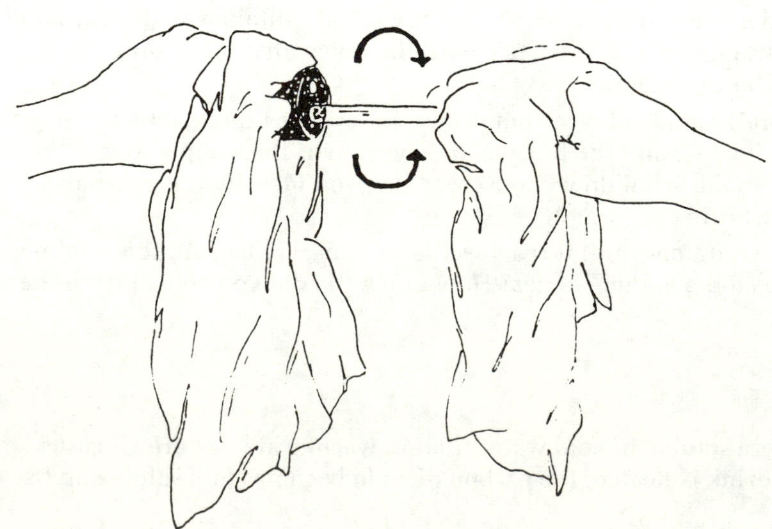

1. You will need a large container or bottle (0.5–1 L) to trap the gas. If you use large graduated cylinders (500 mL), use with care.
2. Club soda works very well.
3. You might also have students determine the amount of air dissolved in tap water by the same procedure. Point out that dissolved oxygen allows fish to live and that heated river and lake water (thermal pollution) reduces the amount of oxygen available to fish.
4. Carbon dioxide dissolved in water forms carbonic acid:

$$H_2O(\ell) + CO_2(g) \longrightarrow H_2CO_3(aq)$$

5. Carbon dioxide is produced and placed in the atmosphere by processes such as the decay of plant and animal matter, sugar fermentation, the burning of carbon-containing fuels, and volcanic eruptions.
6. The oceans dissolve much of the carbon dioxide in the air and thus have kept the amount of this gas in the atmosphere almost constant. Recently, however, scientists have noted that the amount of CO_2 increases every year.
7. You might prefer to determine the amount of carbon dioxide from a large 3-L container of a carbonated beverage as a class demonstration.
8. You may want to analyze the degassed drinks for phosphate content. If so, see Activity 81.

ANSWERS TO THE QUESTIONS

1. The amount will vary, depending upon the temperature and the size of the bottle. However, you can expect about 400 mL of gas per 12-oz bottle.
2. The gas was proven to be carbon dioxide because it reacted with limewater to produce the milky white solid, calcium carbonate. Carbon dioxide in breath produces the same reaction.
3. The solubility of a gas in a liquid decreases as the temperature increases.

3. Solution of a Gas: What Happens During a Pressure Change?

When a gas is dissolved in water, such as carbon dioxide gas in soda pop or air in your blood, the greater the pressure on the gas, the more it will dissolve. The lower the pressure on the gas, the less it will dissolve. You will use a veterinarian's syringe to obtain some club soda and reduce the pressure in the syringe to find out what happens to the carbon dioxide gas in the soda.

Materials

1. Plastic syringe, 60 mL.
2. Club soda.
3. Indicator; methyl red, bromophenol blue, universal indicator, or red cabbage juice.
4. Small bit of putty or clay.

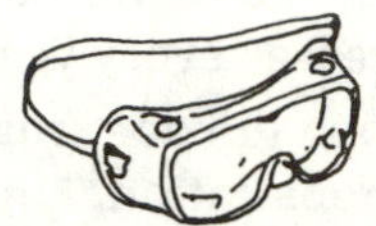

Procedure

1. Place some club soda in a beaker. Add 2 drops of methyl red indicator. Methyl red becomes yellow when the pH is greater than 6 (>6).
2. Fill the syringe one-half full with club soda by putting the open end in the soda and pulling the plunger out.

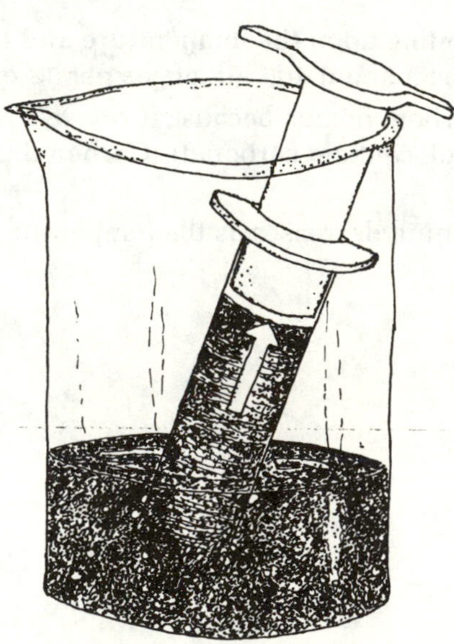

3. Turn the syringe so the open end is up. Gently push the plunger to remove air.
4. Hold a small wad of plastic clay over the open end.
5. Pull the plunger out with force to reduce the pressure in the syringe.
6. Record your observations.

Reactions

1. Carbon dioxide in water produces a weakly acidic solution of carbonic acid:

$$CO_2(g) + H_2O(\ell) \longrightarrow H_2CO_3(aq)$$

2. When the pressure of gases above the solution is less than the pressure of gases in the solution, the carbon dioxide escapes, and the solution becomes less acidic. The previous reaction is reversed:

$$H_2CO_3(aq) \longrightarrow CO_2(g) + H_2O(\ell)$$

Questions

1. Why do you think the liquid goes into the syringe when the plunger is pulled out?
2. In the syringe, how is the pressure above the solution reduced?
3. How do you know the carbon dioxide escaped from the soda?
4. How do you know the escaping carbon dioxide caused the acidity of the soda to be reduced?
5. Why is pulling the plunger out of a sealed syringe difficult?

Notes for the Teacher

BACKGROUND

This experiment is a simple activity that allows the student to explore the ideas of gas pressure and solubility of a gas in a liquid. Syringes full of club soda can be quite intriguing. You may want to demonstrate this one, although letting each student have a chance to try it is worthwhile.

TEACHING TIPS

1. When soda is allowed to stand open, especially in the heat, it goes "flat" because the carbon dioxide escapes. Thus, you know from experience that reduced pressure and/or increased temperature will reduce the solubility of gases in a liquid.
2. The work required to pull the sealed plunger against air pressure is a dramatic example of the effects of air pressure.
3. Try different indicators and different soda solutions.
4. See Activities 33, 47, and 66 for suggested indicators.
5. Plastic syringes can be obtained from a chemical supply house or from a local veterinarian.

ANSWERS TO THE QUESTIONS

1. The pressure of the air on the surface of the liquid in the beaker forces the liquid into the syringe. The pressure in the syringe is low because of the absence of air when the plunger is pulled out.
2. By pulling the plunger.
3. Bubbling is seen.
4. The indicator changed color to show reduction of acidity.
5. Air pressure is a force that acts against the plunger in the sealed syringe. Air pressure is about 15 $lb/in.^2$.

4. Density of Carbon Dioxide

We will investigate a reaction that makes carbon dioxide in your stomach when an antacid is taken for stomach distress. We will let the gas escape from the reaction tube and measure its volume after collecting the carbon dioxide by water displacement. You will also find the mass and calculate the density of carbon dioxide.

Materials

1. Antacid tablet (e.g., Alka-Seltzer) broken into two to four pieces.
2. Test tube, 18 × 150 mm.
3. Stopper, glass bend, and rubber tubing to fit.
4. Tub of water for water displacement of gas.
5. Flask or bottle to collect the gas, 250 mL.
6. Graduated cylinder, 100 mL.
7. Balance sensitive to 0.1 or 0.01 g.

Procedure

1. Place 10 mL of water in the test tube.
2. Weigh the test tube of water and the tablet pieces together. Keep the tablet pieces dry. The test tube may be supported in a plastic cup.
3. Fill the flask or bottle with water. Cover the mouth of the bottle and turn it upside down in the tub of water. Put in the gas delivery tube so that the end of the tube is at the top of the inverted bottle of water.

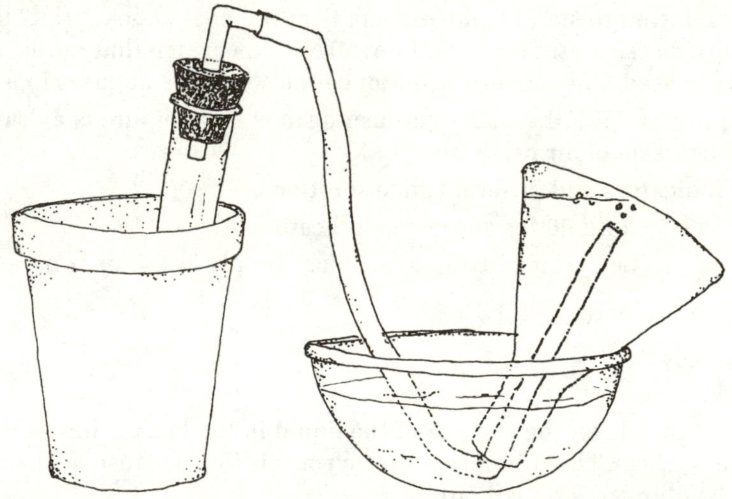

4. Drop the tablet pieces into the test tube and immediately put in the stopper that is attached to the glass bend and the rubber tubing.
5. Collect the gas in the inverted flask or bottle by displacing the water.
6. Measure the volume of the gas by subtracting the volume of the water left in the bottle from the total volume of the bottle.
7. Find the mass of the test tube and contents as before.
8. Determine the change in mass of the tube, cup, and contents that occurred during the reaction. Is this difference the mass of the escaped carbon dioxide?
9. Find the density of the gas (mass/volume).

Reaction

This antacid is a mixture of a powdered base (sodium bicarbonate) and a powdered acid (citric acid). Aspirin (acetylsalicylic acid) is also present. When these substances come into contact in water, they react to produce carbon dioxide, water, sodium citrate, and sodium acetylsalicylate in solution:

$$\underset{}{NaHCO_3(aq)} + \underset{\text{citric acid}}{H_3C_6H_5O_7(aq)} \longrightarrow H_2O(\ell) + CO_2(g) + \underset{\text{sodium citrate}}{NaH_2C_6H_5O_7(aq)}$$

The powdered base will also react with excess stomach acid to reduce the amount of acid in the stomach.

The carbon dioxide bubbles show you that something is happening. They also help to agitate the other gases trapped in the stomach and thus aid in their release.

Questions

1. Describe how a gas is collected by water displacement.
2. What is the volume of the carbon dioxide that you collected?
3. How was the mass of the gas determined?
4. What was the mass of carbon dioxide produced by your antacid tablet?
5. What is meant by "density"?
6. What is the density of carbon dioxide at room temperature and normal room pressure?
7. Compare your results with those of the other students in your class.

Notes for the Teacher

BACKGROUND

Equal volumes of gases have equal numbers of molecules, but different molecules may have different masses. Therefore, the density of each gas will be unique to that gas. Students will have fun finding out that gases have masses and that the mass of a gas can be measured. This activity with a very common substance should help students gain skills, including the collection of a gas by water displacement, the measurement of gas volume, and the determination of density. They may also appreciate the chemical reaction observed when this antacid is dropped into water.

TEACHING TIPS

SAFETY NOTE: It is best to prepare the stopper and tubing assembly for the students. Fire polish each end of the glass tubing. (Heat it carefully in a hot flame until the sharp ends are smooth.) Lubricate the tubing with glycerol or water and CAREFULLY insert the tubing in the stopper with a twisting motion. You may make a bend before inserting the glass in the stopper.

1. This activity works best with the antacid Alka-Seltzer.
2. Two students are needed to do this activity. One will need to keep a good hold on the inverted flask or bottle while the other adds the tablet pieces to the tube.
3. You could ask students to try to figure out a reasonable procedure for this activity, as a group, before you give them explicit directions.
4. Small pails, buckets, and dishpans make good water tubs.
5. If students compare results, you can discuss reproducibility and precision.

6. The formation of gas is complete in about 10 min.
7. Other gas production reactions could be performed to compare densities. Try granular zinc plus 1 M hydrochloric acid or 3% hydrogen peroxide (store variety) plus powdered manganese dioxide or potassium iodide.
8. If your balances are not sensitive to 0.1 or 0.01 g, this works well as a demonstration with 10 tablets in a 250-mL flask, with a large bottle to collect the gas.

ANSWERS TO THE QUESTIONS

1. See Procedures 3–5.
2. Between 100 and 200 mL, depending upon how fresh the antacid is.
3. See Procedures 2, 7, and 8.
4. The mass is the difference between the initial mass and the final mass.
5. *Density* is the ratio of mass to volume (mass/volume).
6. About 2 g/L or 0.002 g/cm^3.
7. Results should be about the same.

5. Behavior of Gases and the Boiling Egg

Several fundamental laws help us to describe the behavior of gases. Charles's law tells us that an increase in temperature will cause an increase in the volume of a gas if the pressure does not change. Boyle's law states that an increase in pressure on a gas will cause a decrease in volume, if the temperature remains constant. Another gas law, proposed by Guillaume Amonton in 1703, tells us that increasing the temperature of a gas will increase its pressure if the volume remains constant.

In this activity we will increase the temperature of a fixed volume of air trapped in an egg and note the effect of the increasing pressure.

Materials

1. Two raw eggs.
2. Large beaker.
3. Burner.

Procedure

1. The beaker should be large enough for two eggs to fit inside easily.
2. Fill the beaker two-thirds full with water and heat until it is near boiling.
3. Using a pencil, mark one egg with a large "X". Carefully punch a small hole in the large end of this marked egg with a straight pin. Twist the pin as you make the hole to avoid breaking the shell.
4. Do nothing to the other egg; it will serve as a control.
5. Carefully place both eggs in the beaker of hot water.
6. Observe both eggs closely. Do you see tiny air bubbles forming around the control egg? Is a steady stream of bubbles coming from the hole in the marked egg? What do you think is happening?

Reaction

The eggshell is slightly *porous*, which means that it has many small holes. When the air inside the egg is heated, the pressure of the gas increases, and some of the air is forced through these tiny pores. What would happen if the egg did not contain these pores? The pressure would cause the eggshell to crack and break. Sometimes breakage occurs anyway when the temperature is rapidly increased, for example, by placing the egg immediately in boiling water.

If we place a hole in the shell, the increased pressure of the air causes the expanding volume of gas to leave the shell faster by forming a stream or jet of air bubbles. This situation allows us to observe direct evidence that increasing the temperature of the gas in the egg results in an increase in gas pressure.

Questions

1. State Amonton's law.
2. Can you think of other examples of Amonton's law?

Notes for the Teacher

BACKGROUND

This activity gives students experience with a very important gas law. Amonton determined that an increase in temperature causes an increase in gas pressure, if the volume of gas remains the same. We will experiment with the fixed volume of gas trapped inside an egg. According to Amonton's law, when this fixed volume of air is heated, the pressure should increase. We can see that the pressure does increase because it forces the air out through the tiny pores in the eggshell. If a hole is placed in the egg, the increasing pressure causes a stream of gas bubbles. Actually, the volume of gas does not remain constant. If it did remain constant, the egg would crack and break (as it sometimes does) when the temperature is suddenly increased.

TEACHING TIPS

1. Sometimes when an egg is boiled quickly, the protein coagulates before all of the air can escape through the pores in the shell. This occurrence gives the boiled egg a blunt end rather than a rounded end.
2. Students should not eat their experiment.
3. Ask students why placing the egg in lukewarm water for a few minutes before boiling often prevents the shell from cracking.
4. The sizes of pores in eggs vary. Some eggs have large pores, and other eggs have very small pores.
5. Some students may confuse Amonton's law with Charles's law (which is sometimes called Gay-Lussac's law because Joseph Gay-Lussac expanded Charles's law in 1802). Charles's law does not apply here because the pressure is not held constant. In fact, the increase in pressure causes the bubbles of air to form outside the shell.

ANSWERS TO THE QUESTIONS

1. Increasing the temperature of a gas results in an increase in pressure, if the volume of the gas remains constant.
2. Other examples include many explosions (the production and expansion of gases in a confined area cause a sudden increase in pressure), a bullet fired from a gun (the increased temperature from the ignition of the gun powder causes an increase in pressure that forces the bullet from the shell), the popping of popcorn (the increased temperature vaporizes a small amount of water in the kernel and thus increases the pressure and causes the kernel to pop), and the "spewing" of bottled soft drinks when opened on a hot day.

6. Used Breath: Carbon Dioxide in Exhaled Air

Your exhaled breath is a mixture of gases from your lungs. Some of the gases go into your lungs from your blood. One of the gases that comes from the blood is carbon dioxide (CO_2), which is produced throughout the body when glucose ($C_6H_{12}O_6$) and oxygen (O_2) react to form water (H_2O), carbon dioxide, and energy. You will study some of the properties of carbon dioxide by exhaling through a straw into a solution of a base, calcium hydroxide [$Ca(OH)_2$]. Calcium hydroxide solution is known as limewater.

Materials

1. Limewater [saturated calcium hydroxide ($Ca(OH)_2$) solution].
2. Phenolphthalein, an indicator of a basic solution when pink.
3. Drinking straw.
4. Test tube.
5. Square of aluminum foil to fit over the mouth of the test tube.

Procedure

1. Place 10–15 mL of calcium hydroxide solution in a large test tube.
2. Add 2 drops of phenolphthalein solution. Observe.
3. Seal the tube with the square of foil. Carefully push the straw through the foil or make a small hole first with a pencil or pin.
4. Blow GENTLY through the straw into the tube of calcium hydroxide solution. Be careful. Calcium hydroxide is a base.
5. Blow gently until you see three changes take place. Describe the changes.

Reactions

1. Respiration reaction, which produces CO_2 in cells:

$$C_6H_{12}O_6(s) + 6O_2(g) \longrightarrow 6CO_2(g) + 6H_2O(\ell) + \text{energy}$$

2. Acid-forming reaction:

$$CO_2(g) + H_2O(\ell) \longrightarrow H_2CO_3(aq)$$

3. Neutralization reaction (acid added to base):

$$\underset{\text{acid}}{H_2CO_3(aq)} + \underset{\text{base}}{Ca(OH)_2(aq)} \longrightarrow 2H_2O(\ell) + \underset{\text{cloudy precipitate}}{CaCO_3(s)}$$

The excess acid formed by the carbon dioxide reacts with the calcium carbonate to dissolve it. Calcium carbonate is in limestone, marble, and eggshell.

Questions

1. Describe the first change you saw.
2. Describe the second change.
3. Describe the third change.
4. How did you know that an acid was being added to the tube as you blew?
5. Calcium carbonate caused the cloudiness of the solution. The calcium was already in the calcium hydroxide solution. Where do you think the carbonate came from?

6. What do you think reacted with the calcium carbonate to make the solution clear again?
7. State one thing you learned about the carbon dioxide in your breath.

Notes for the Teacher

BACKGROUND

This activity helps students to connect themselves with the material world and to appreciate the chemical balance within them. Develop the chemistry of the reactions to whatever extent your students are capable or interested.

TEACHING TIPS

SAFETY NOTE: Calcium hydroxide is a base. Students must wear goggles.

1. Make saturated calcium hydroxide by putting several teaspoons of either calcium hydroxide or calcium oxide in a jar of water until some solid remains on the bottom. Cap the jar and let the solution settle overnight. Pour off the clear saturated solution into the test tubes.
2. If your students do not need volume-measuring practice, pour the solution into the test tubes for them. The amount is not important.
3. Phenolphthalein is a substance produced by a fungus that is pink in slightly basic solution. It can be obtained from chemical supply sources or by soaking a powdered laxative (e.g., Ex-Lax) in alcohol.
4. Be sure the students blow gently so that they do not blow solution into their faces. Use long straws and large test tubes covered with foil.
5. Chemists use the symbol (s) to mean a solid, (ℓ) to mean a liquid, (g) to mean a gas; and (aq) to mean an aqueous solution.

ANSWERS TO THE QUESTIONS

1, 2, and 3. The solution will become colorless (acid from CO_2 neutralizes the base), then cloudy (calcium carbonate forms), and then clear again (acid reacts with the calcium carbonate).

4. Pink base indicator turned colorless.
5. From the CO_2.
6. Acid from CO_2 in water.
7. Student answers will vary.

7. Reactions of Eggshells, Seashells, and Baking Soda

Carbon dioxide (CO_2) is an odorless, colorless gas that is produced by compounds that contain the carbonate ion (CO_3^{2-}). This gas is also exhaled by humans and other animals and given off when wood and gasoline burn. In this activity you will investigate compounds containing carbonate that react with acids to release carbon dioxide. Using a known acid, we can identify carbonates with this reaction. Using a known carbonate, we can also identify acids with this reaction.

Materials

1. Calcium carbonate ($CaCO_3$).
2. Baking soda, sodium bicarbonate ($NaHCO_3$).
3. Snail, clam, or oyster shells.
4. Eggshells.
5. Limestone.
6. Marble chips or decorative marble, unpolished.
7. Any non-carbonate such as gypsum dry wall, table salt, or calcium chloride road salt.
8. Hydrochloric acid (HCl), 0.1 M.
9. Vinegar (dilute acetic acid).
10. Lemon juice.
11. Club soda (Seltzer water).
12. Any non-acid such as ammonia or soapy water.
13. Limewater (saturated calcium hydroxide, $Ca(OH)_2$, solution) with droppers.
14. Microscope slides.
15. Test tubes.

Procedure

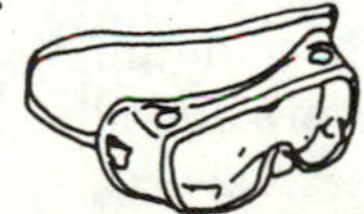

1. Place a small amount of calcium carbonate in the bottom of a small test tube.
2. Add several drops of hydrochloric acid to the tube. Observe the fizzing, which indicates the formation of a gas.
3. Prepare a test drop of limewater (calcium hydroxide solution) by placing a drop on a clean dry microscope slide.
4. Repeat Procedures 1 and 2, then immediately turn the prepared slide over the mouth of the test tube so that the drop of calcium hydroxide solution is hanging upside down.
5. Look for the drop to turn white or cloudy.
6. Place the other solid samples in separate test tubes. Repeat the "hanging-drop test" for carbon dioxide for each sample by adding a few drops of hydrochloric acid to each test tube and hanging a drop of limewater over the mouth of each test tube.
7. Make a table of your results.
8. Using laboratory-grade calcium carbonate or any one of the other carbonates, test for the presence of acid in each of the solutions by using the same hanging-drop test.
9. Make a table of your results.

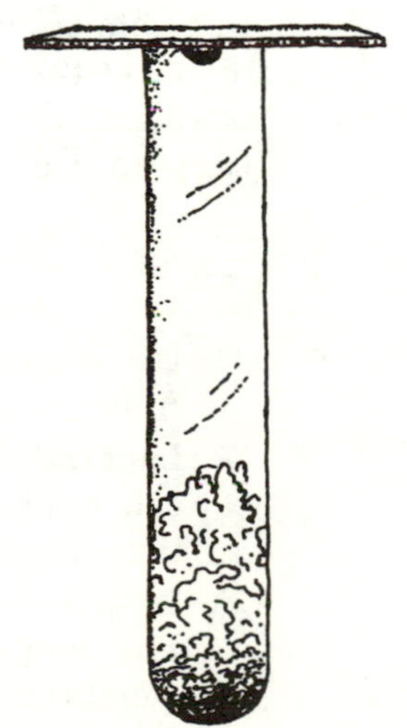

Reactions

1. Any acid added to a carbonate gives carbon dioxide gas, water, and a salt:

$$\underset{\text{carbonate}}{CaCO_3(s)} + \underset{\text{acid}}{2HCl(aq)} \longrightarrow \underset{\text{carbon dioxide}}{CO_2(g)} + \underset{\text{water}}{H_2O(\ell)} + \underset{\text{salt}}{CaCl_2(aq)}$$

2. When the carbon dioxide gas reaches the hanging drop of calcium hydroxide solution, calcium carbonate forms in the drop, and the drop turns white:

$$CO_2(g) + Ca(OH)_2(aq) \longrightarrow \underset{\text{white}}{CaCO_3(s)} + H_2O(\ell)$$

Questions

1. Did all of the solids produce a gas when the known acid was added?
2. Did all of the solutions react like acids with the known carbonate?
3. What is the white precipitate that causes the drop of calcium hydroxide to turn white?
4. Why must you be careful not to breathe on the drop of calcium hydroxide?
5. Describe how you would use this test to identify an acid.

Notes for the Teacher

BACKGROUND

This activity allows students to practice good scientific process and to learn some simple acid–base chemistry. Reactions between acids and carbonates are quite common: Acid rain is produced when sulfur dioxide reacts with water. The acid rain reacts with carbonates in statues to remove the surface detail. Carbonates are added to acidic lakes to reduce the acidity caused by acid rain. Carbonates in the soil, often from fossilized seashells, react with slightly acidic normal rainfall to form underground caves. Geologists use the acid test to identify limestone. Snail population data are used as indicators of acidic pollution in streams because the acid dissolves the shells of snails. Thus, few snails are found in acidic environments. In baking, acids reacting with baking soda in batters cause the "quick" breads and cakes to rise because of the production of carbon dioxide gas.

TEACHING TIPS

1. Use any compounds of carbonate and any acids you have available. See Appendix 2 for dilution of HCl.
2. Limewater is prepared by adding 1 heaping tablespoon of calcium oxide or calcium hydroxide to 1 L of water to form a saturated solution. Stir well. Allow the undissolved solid to settle overnight and pour off the clear solution.
3. Try placing a whole uncooked egg in an opened 1-qt jar of vinegar overnight. The shell, which is calcium carbonate, will react with the acid, and the inner membrane will be left intact. Students are amazed by this result. This activity is a great one to do at home.
4. This experiment reinforces the idea of controlled variables and predictability of carbonate reactions as well as acid reactions.

5. You could introduce the ideas of solubility and acid–base reactions with this activity. When carbon dioxide reacts with calcium hydroxide, it acts as an acid. The calcium hydroxide is a base.
6. Barium hydroxide [$Ba(OH)_2$] solution is more sensitive than calcium hydroxide solution in this test; however, $Ba(OH)_2$ is toxic and should not be used.
7. You might try blackboard chalk; however, most blackboard chalk is actually calcium sulfate and not calcium carbonate.
8. Dry wall is calcium sulfate. Road salt is calcium chloride.

ANSWERS TO THE QUESTIONS

1. No, only the carbonates.
2. No, only the acids. Ammonia is a base.
3. Calcium carbonate ($CaCO_3$).
4. Carbon dioxide in exhaled breath will cause a precipitate of calcium carbonate to form.
5. The reaction of any acid with a pure carbonate such as calcium carbonate or sodium bicarbonate should produce bubbles of carbon dioxide.

8. What Makes Popcorn "Pop"?

If we examine a single kernel of popcorn and compare it with a grain of ordinary corn, two differences can be easily noted. First, the kernel of popcorn is more rounded than that of ordinary corn. Second, the coat on the popcorn is thicker and tougher. Both of these characteristics are important in the popping of corn. You will investigate the conditions and results of corn popping.

Materials

1. Popcorn.
2. Ordinary corn.
3. Cooking oil.
4. Flask and stopper.
5. Burner.
6. Pin.
7. Tongs or a pot holder.

Procedure

1. Examine the popcorn and the ordinary corn. Describe as many differences and similarities as you can.
2. Select two samples of popcorn, each containing 10 kernels. Determine the mass of each sample of 10 kernels and record. Calculate the average mass of one kernel in each sample and record.
3. Coat the bottom of the flask with cooking oil. Pour out the excess. Place one sample of popcorn kernels in the flask and gently heat. Hold the flask with tongs or a pot holder. Shake the flask. Continue until the popcorn pops.
4. Determine the mass of the sample and the average mass of a single popped kernel.
5. Treat the second sample as you did the first. However, before heating, carefully pierce the seed coat of each kernel with a pin.

Reaction

How can we account for the loss of mass of the popped kernel? The loss represents water, which has escaped "explosively" from the kernel as steam:

$$H_2O(\ell) \longrightarrow H_2O(g)$$

The white starch grains "fluff up" as the trillions of water molecules fly out of the kernel and burst through the seed coat.

Questions

1. Compare the average mass of a popped kernel with that of an unpopped kernel. If 18 g of water has 600,000,000,000,000,000,000,000 molecules of water (i.e., 6×10^{23} molecules), how many molecules of water escaped from one average kernel?
2. What happened to the popcorn that was pierced before heating?

Notes for the Teacher

BACKGROUND

Even though popcorn seems to be dry, each kernel contains a small amount of water. When the popcorn is heated, the water forms steam. When sufficient pressure builds up within a kernel, the thick coat bursts to allow the steam to escape. This activity is a good way to introduce students to balances, heat sources, and the collection and analysis of data.

TEACHING TIPS

1. Encourage students to create a useful and well-organized data table. Pooling the ideas of the class about how data might be arranged is helpful. The following is a suggested model:

	Test 1	Test 2
Mass of 10 unpopped kernels		
Average mass of a single kernel		
Mass of 10 popped kernels		
Average mass of a single kernel		

2. Emphasize recording data and averaging to obtain the mass of a single kernel of corn. Focus on the limitations of averaging.
3. Students can extend this activity to further investigation by comparing the steam loss, popping rate, volume, and other characteristics of various brands and types of popcorn.
4. In Procedure 5, piercing the kernel allows the steam to escape and prevents pressure from building up and exploding the kernel; thus the popcorn does not pop.
5. Do not eat the popcorn.

ANSWERS TO THE QUESTIONS

1. If the average water loss per kernel is 0.18 g, then the average water loss per kernel amounts to 6×10^{21} molecules. By considering the molecules trapped between the starch grains and cellulose of the kernel, you can imagine how the fluffing action occurs when the molecules escape.
2. Pierced popcorn will not pop.

9. Sublimation of Air Freshener

Sublimation is one of the most interesting physical changes. When a substance *sublimes*, it goes directly from a solid to a gas without passing through the liquid state. Dry ice sublimes, as do iodine and mothballs (the vapor from the mothball keeps away the moth). In this activity we will study another common substance that sublimes: air freshener.

Materials

1. Small piece of solid air freshener.
2. Two beakers, 100 and 150 mL.
3. Shallow dish or pan.
4. Ice.
5. Thermometer.

Procedure

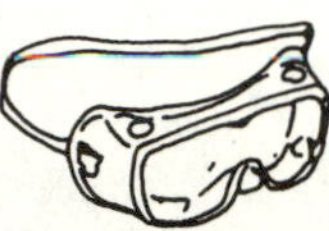

1. Place a few small lumps of air freshener in the bottom of the 150-mL beaker.
2. Put the 100-mL beaker inside the 150-mL beaker. Notice that it fits nicely but does not reach the bottom of the 150-mL beaker. If you use a toilet bowl freshener, do this activity in the fume hood.
3. Fill the small beaker three-fourths full with ice. BE SURE THAT NO ICE GETS INTO THE LARGER BEAKER.
4. Fill the shallow dish or pan about one-third full with hot water.
5. Measure the temperature of the water bath and adjust it by adding cold water until the temperature of the water bath is about 45 °C.
6. Place the sublimation apparatus in the shallow dish.
7. Observe what happens to the solid.

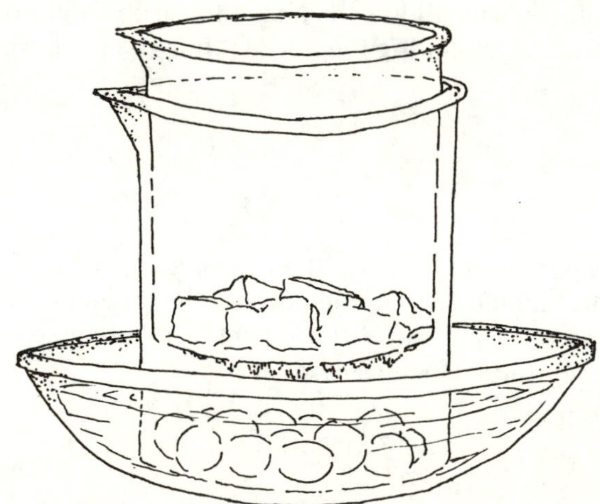

Reaction

The heat from the water bath causes the solid air freshener to vaporize (sublime). The cold smaller beaker causes the vaporized air freshener to condense and re-form the solid.

Questions

1. Do you think that there is anything special about 45 °C? Try lower temperatures and higher temperatures.
2. Do you think that this activity would work with mothballs or solid iodine? Try it, but use only small amounts in a hood.
3. Define “sublimation”.

Notes for the Teacher

BACKGROUND

Sublimation is, strictly speaking, the vaporization of a solid. The opposite process, the formation of a solid directly from a vapor, is called *deposition*. Sometimes the term “sublimation” is used for both processes.

Mothballs sublime. Those that contain camphor will sublime at room temperature. Sometimes you can find crystals of camphor on clothes that have been stored over mothballs. Naphthalene, another substance often used in mothballs, also sublimes.

TEACHING TIPS

1. You can use other materials that sublime, but solid toilet-bowl cleaners (the kind with a wire hook) work best. If you use iodine, use only a few crystals and do the activity in a hood. Iodine is sufficiently toxic to require this procedure.
2. Naphthalene mothballs must be heated to near 70 °C for sublimation to occur.
3. Dry ice sublimes at −78.5 °C and above.
4. If possible, use colored air freshener; notice that the material that collects on the cold beaker is white! The dye does not sublime because it is not chemically a part of the compound that does sublime.
5. Vapor deposition is an important industrial process for separation and purification.
6. If you use cheap toilet-bowl deodorizers that contain *p*-dichlorobenzene (also found in some types of mothballs), handle them with tongs in the hood. *p*-Dichlorobenzene is toxic.
7. Be patient. This activity may take a while.
8. If a thermometer breaks, take special precautions. Collect and store mercury in a labeled and sealed container. Use a commercial mercury spill collection kit. See Appendix 4B.

ANSWERS TO THE QUESTIONS

1. Yes, the air freshener will not sublime measurably below this temperature.
2. The activity will work with a variety of substances that sublime. See previous suggestions.
3. *Sublimation* is the physical change that occurs when a substance goes from a solid phase directly to a gaseous phase.

Chemistry of Matter: Liquids and Solids

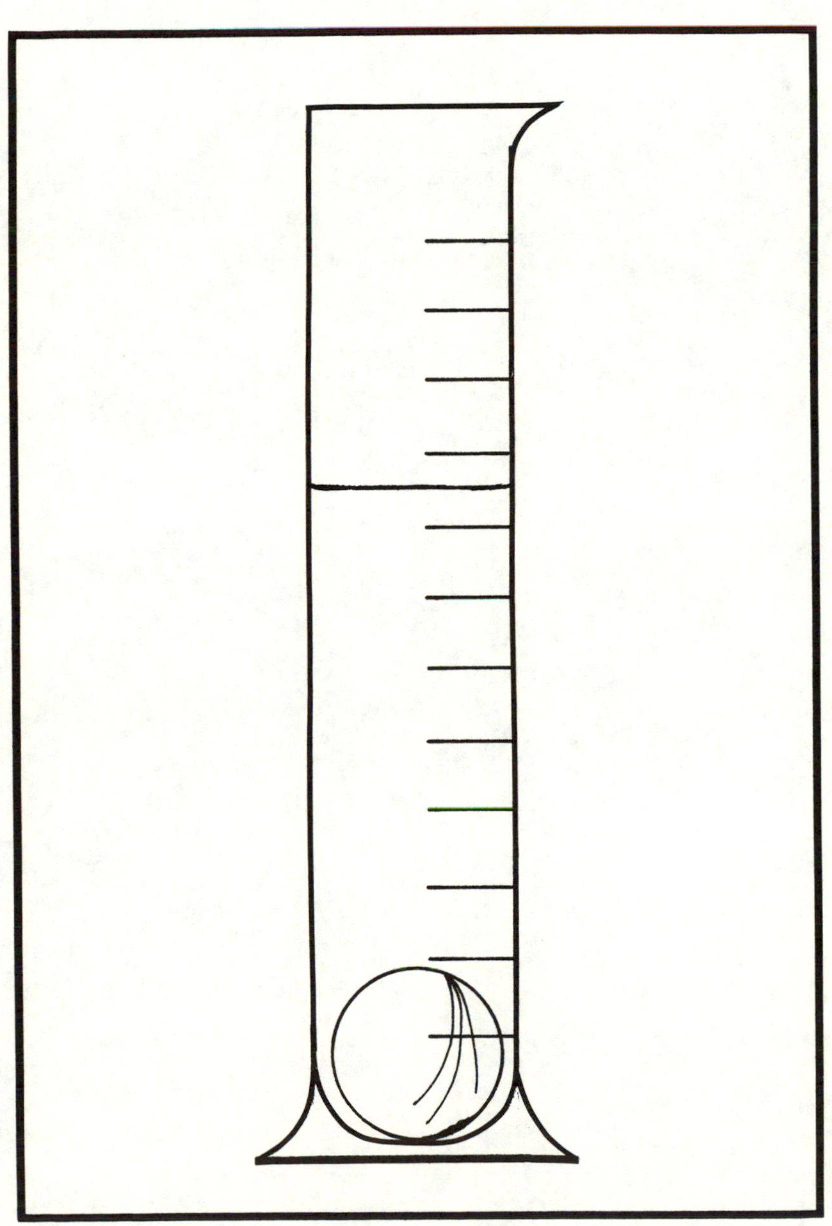

10. Layers of Liquids

Pure liquid substances or solutions may be different from each other in many ways. One difference might be in how heavy equal volumes of each liquid are. You can study this difference by adding four liquids in layers to a cylinder. Those with greater mass per volume will tend to remain below those with less mass per volume. *Density* is the mass per volume.

Materials

1. Graduated cylinder, 50 or 100 mL.
2. 10 mL of each of the following: sugar water (corn syrup), strong cold coffee, green alcohol (use food coloring), and vegetable oil.
3. Small objects of varying densities.

Procedure

1. Pour the sugar water (corn syrup) into the cylinder to a depth of 10 mL.
2. Slowly add 10 mL of cold coffee on top of the syrup by pouring it down the side of the cylinder.
3. Slightly tip the cylinder and add 10 mL of vegetable oil on top of the coffee.
4. Carefully add 10 mL of colored alcohol on top of the oil. Alcohol is flammable. Extinguish all flames in the room.
5. Set the cylinder where it will not be disturbed and where you can observe it each day.
6. Find some small objects, for example, marbles, plastic pieces, and paper clips. Predict on which layer each object will float and then drop each object into the cylinder.
7. Try adding the liquids to the cylinder in a different order to see if they make layers in the same way.

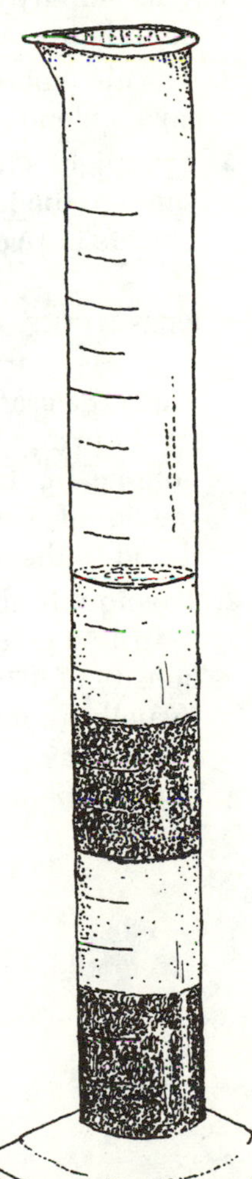

Reaction

Two characteristic properties of substances are being observed in this activity:

1. Some liquids are more dense than others. Those that are more dense will rest in the cylinder below those that are less dense.
2. Some liquids cannot be mixed with other liquids. We say they are *immiscible.*

These two properties of the liquids tend to keep them separate from each other in the cylinder. After several weeks, however, slow movement of the molecules will cause the layers to mix.

Questions

1. How could you measure the density of each liquid?
2. If water has a density of 1.00 g/cm^3, suggest a liquid that probably has a density less than 1.00 g/cm^3.
3. Name some liquids that will not dissolve in water.

Notes for the Teacher

BACKGROUND

Density is an idea that many students find difficult to grasp because it depends on two measured quantities, mass and volume, which are set in a ratio—mass/volume. An important step in student reasoning is to develop proficiency with ratios. This activity allows the student to practice comparing densities in a very concrete way and does not require the use of numerical symbols at all.

Solubility is another idea that the students will consider as they notice the separation of the liquids. Some substances are composed of molecules that have properties that repel the molecules of another substance. The two are therefore *insoluble* or *immiscible*.

TEACHING TIPS

1. Use green food coloring to color the alcohol. Limit the amount of alcohol that you dispense. A good rule is *no more than 250 mL of a flammable liquid per dispensing container.* Use one or two dispensing containers.
2. This activity could be performed for the class in a large graduated cylinder (250 or 500 mL) to allow the class as a whole to discuss their observations.
3. This activity could also be done on a smaller scale with a small graduated cylinder or test tube.
4. The alcohol is colored so that its layer will be more visible, especially as the layers blend slowly over several days or weeks.
5. You might save each team's density column for a different purpose: Stir one. (Do some of the liquids mix with others when stirred?) Keep one undisturbed. Place objects in two other ones. Add water or another liquid to another one.

ANSWERS TO THE QUESTIONS

1. Place the graduated cylinder on a balance. Take a reading of mass. Add the liquid to any level. Take a reading of the mass again. Take a reading of the volume. Subtract the mass of the cylinder from the mass of the cylinder with the liquid. The result is the mass of the liquid. Find the ratio of the mass of the liquid to the volume of the liquid.
2. The liquids above the coffee (which is mostly water) would have a density less than 1.00 g/cm^3. Most simple alcohols (e.g., propanol and ethanol) have a density of 0.79 g/cm^3. You might have students find the density of vegetable oil, (usually about 0.9 g/cm^3). Finding the density of water is also a useful experience.
3. Oil, gasoline, and turpentine.

11. Pennies in the Glass and Properties of Water

In this activity we will examine one of the fundamental physical properties of water, namely, its surface tension. *Surface tension* is the force that makes the surface of a liquid act as if it were a stretched membrane. Because of this effect, a glass full of water will not overflow when several pennies are dropped into it. See how many pennies you can add to the glass before it overflows!

Materials

1. Large drinking glasses.
2. Pennies (about 50 per group).

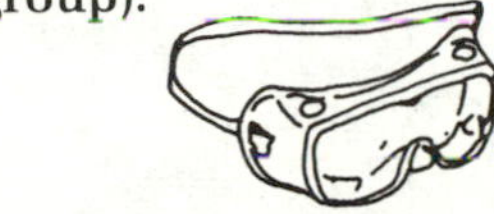

Procedure

1. Carefully fill your glass or tumbler with water until it is full but not overflowing.
2. Carefully drop a penny into the glass of water. Hold the penny just above the surface of the water and drop it edge down in the center of the glass.
3. Continue dropping the pennies, one at a time, into the glass until it overflows. How many pennies were you able to add? What did the surface of the water look like just before it overflowed?
4. Try the same procedure with rubbing alcohol or a soap solution instead of water. Remember, rubbing alcohol is flammable. Extinguish all flames in the room.

Reaction

Because of surface tension, water molecules on the surface of the water are attracted to each other in an attempt to pull the water molecules back into the liquid. This attraction is so strong that the water will actually form a curved surface above the glass (convex surface) as pennies are added to displace small volumes of water.

Questions

1. How many pennies could you drop into the glass before it overflowed?
2. How can you explain this phenomenon?
3. Do you think that other liquids, such as alcohol, would behave the same way? Try it!
4. How many drops of water can you place on a penny before the water overflows?

Notes for the Teacher

BACKGROUND

This activity is a nice way to illustrate the surface tension of a liquid. Students are often surprised to find that they can drop 25–40 pennies into a glass full of water before the water overflows. *Surface tension* is the force that results when molecules on the surface of a liquid are pulled in and down by the attraction of other water molecules. This force causes the surface of the water to act as if it were a stretched membrane.

TEACHING TIPS

1. Molecules within a liquid are attracted equally in all directions by other neighboring molecules:

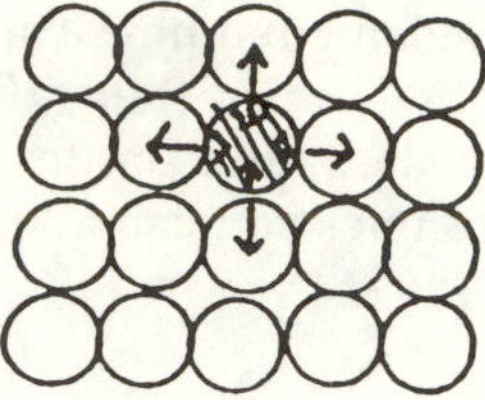

However, water molecules on the surface of the liquid are attracted only into the liquid and to the sides:

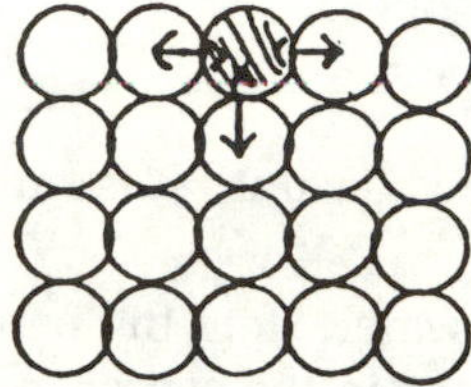

This unbalanced force tends to pull the water molecules back into the liquid and thus forms the convex surface.

2. Liquid drops are spherical because of their surface tension. In a sphere, the ratio of surface area to volume is at a minimum.
3. Some insects such as the water strider can walk on the surface of water, and a needle carefully placed on the surface of water will not sink. Both the water bug and the needle are more dense than water; however, they do not sink because of the surface tension of the water.
4. Surface tension is also one of the forces that causes water to rise in a plant from the roots.
5. Soap bubbles are spherical because of surface tension.
6. You can reduce the surface tension of a liquid with alcohol or soap solution. Float two plastic toothpicks or two needles on the top of a glass of water. If you are careful, you can also succeed with a paper clip or a thumbtack. Add either 1 drop of alcohol or 1 drop of soap solution between the two. They will sink because the added molecules interfere with the attraction of water molecules for each other, and thus the surface tension is lowered.
7. You can also demonstrate lowered surface tension with a rubber band on the surface of water. Notice that the rubber band keeps its normal oblong shape. Now, place a drop of soap solution or alcohol in the middle of the rubber band. The surface tension of the water inside the rubber band will be decreased, and the normal surface tension of the water outside the band will pull on it and cause it to take the shape of a circle.

ANSWERS TO THE QUESTIONS

1. The number of pennies will depend on the size of the tumbler. However, students will be amazed at the large number!
2. As the volume of water is displaced by the penny and pushed up, surface tension acts to pull the water molecules toward the middle and downward. This situation produces a stretched membrane effect on the surface of the water.

3. Alcohol has a lower surface tension than water; therefore, it will not work as well as water. The intermolecular forces pulling alcohol molecules together are weaker than those pulling water molecules together; thus, alcohol will have less of a tendency than water to form a convex surface as pennies are dropped into it. In addition, alcohol has a lower boiling point than water, which also suggests weaker intermolecular forces in the alcohol.
4. Answers will vary; however, the number will be greater than you expect.

12. Boiling Water in a Paper Cup

We often boil water in a metal pan or ceramic container because these do not burn. However, a paper cup will burn. Can you heat water in a paper cup until the water boils? In this experiment we will find out if this result is possible.

Materials

1. Paper cups.
2. Burner or candle.
3. Ring and ring stand or other support.

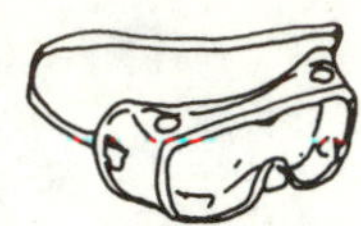

Procedure

1. Select a small paper cup, or make one by folding a sheet of notebook paper.
2. Support the cup by placing it in a small ring clamped to a ring stand or on a tripod.

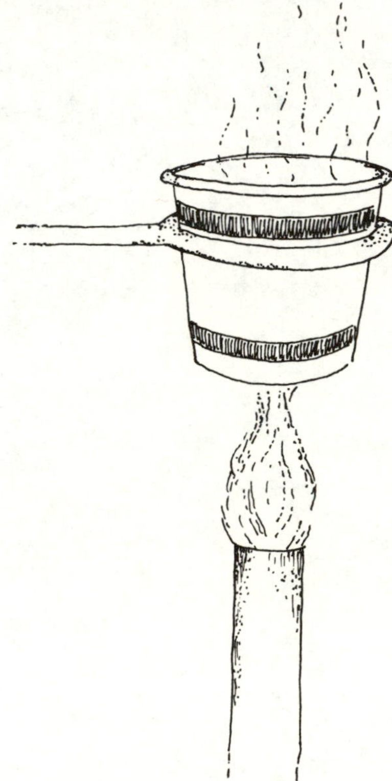

3. Carefully add tap water to the cup until it is filled to the very top but not overflowing.
4. Gently heat the paper cup with the flame from a burner or candle. Be careful with flames and with hot water.
5. Continue heating. Can you make the water boil?
6. Remove the heat and allow the cup and water to cool.

Reaction

Water conducts heat energy away from the paper; therefore, the temperature of the paper does not rise appreciably above that of the water and does not get hot enough to burn.

$$H_2O(\ell) + \text{heat} \longrightarrow H_2O(g)$$

Questions

1. Examine the cup after the water has cooled. What do you observe?
2. Why did the paper cup not burn?
3. If you heat water in a glass container, the glass will get hotter than the water. However, the paper did not get very hot. How can you explain this situation?

Notes for the Teacher

BACKGROUND

When heat energy is absorbed by matter (e.g., glass, paper, and water), the energy is converted to kinetic energy. Temperature is a measure of kinetic energy. Thus, as heat is added to matter, the temperature rises. Some substances conduct heat energy more readily than other substances. Water, for example, conducts heat energy more readily than paper. Therefore, when a paper cup of water is heated, most of the heat is readily transmitted to the water, and little is left to be absorbed by the paper cup. The water continues to absorb heat until it reaches a high enough temperature for the boiling process to occur, and then it continues to absorb energy to change into steam. As long as water is left in the paper cup, the paper will remain well below its combustion point. Thus, the paper does not burn.

TEACHING TIPS

1. Caution students about the use of a burner. Do not allow them to heat poorly constructed paper cups that might come apart when filled with hot water.
2. Dixie Cups, cone-shaped bathroom cups, and cups used to dispense pills in the hospital work well. You cannot use wax-coated cups or cups with large and thick top or bottom edges.
3. You can expect exposed edges of the paper cups to char, but they will not burn.
4. Lead students into a discussion of things that conduct heat readily (e.g., copper, aluminum, and most other metals) and things that do not readily conduct heat (e.g., wood, ceramics, and plastics).
5. Ask students why many stainless steel pans have a coating of copper on the bottom. (Copper conducts heat better than stainless steel.)
6. With care, you can even do this activity with a paper bag as long as no edges are exposed to the flame. Be careful! The bag may break.
7. Try the following as a demonstration: Add about 1 mL of water to a balloon. Now blow into the balloon to partially inflate it. Carefully heat the bottom of the balloon, where the water is pooled, with a lighter. Why doesn't the balloon pop?

ANSWERS TO THE QUESTIONS

1. The paper cup may be slightly charred around the parts not in contact with water, but the cup will not be scorched or burned.
2. The water conducts the heat from the paper, leaving little heat available to burn the paper.
3. Thick glass is a poor conductor when it is used as a container for boiling water. Much of the heat raises the temperature of the glass instead of that of the water.

13. Hydrates: Molecules with Water Attached

Water molecules often combine with other substances with the water molecule remaining intact. This process is called *hydration*, and compounds containing molecules of water as part of their structure are called *hydrates*. When a solution of a hydrated salt is evaporated, the salt that remains is made of water molecules and charged particles (the salt) combined in a definite proportion. Examples include $CuSO_4 \cdot 5H_2O$ (copper sulfate pentahydrate), $Co(NO_3)_2 \cdot 6H_2O$ (cobalt nitrate hexahydrate), and $Na_2SO_4 \cdot 10H_2O$ (sodium sulfate decahydrate). When all of the water is removed, the resulting salt is called the *anhydrous* form.

In this activity, we will add water back to a dehydrated form of calcium sulfate to form the hydrate. We will observe the special properties of this hydrate.

Materials

1. Several coins.
2. Small paper cups.
3. Calcium sulfate hemihydrate (plaster of Paris).
4. Small beaker and stirring rod.

Procedure

1. Place several coins or other small objects that you wish to mold (e.g., buttons and paper clips) in the bottom of a paper cup.
2. Place enough calcium sulfate hemihydrate in a beaker to give a depth of about 1 in.
3. Add a small amount of water—just enough to make a thick soupy paste.
4. Stir thoroughly to break up any large lumps of solid and to produce a uniform paste.
5. Pour the paste into the cup containing the coins or other objects.
6. Set the cup aside for 15–30 min until the paste solidifies.
7. Cut and tear away the paper cup, remove the coins or other objects, and examine the plaster cast that you have produced.

Reaction

Plaster of Paris is formed by heating calcium sulfate dihydrate until it loses three-fourths of its water. In this activity, we add the water back to plaster of Paris to again produce the dihydrate.

$$\underset{\text{calcium sulfate hemihydrate}}{(CaSO_4)_2 \cdot H_2O(s)} + 3H_2O(\ell) \longrightarrow \underset{\text{calcium sulfate dihydrate}}{2CaSO_4 \cdot 2H_2O(s)}$$

Notice that one molecule of water is attached to two molecules of $CaSO_4$ in calcium sulfate hemihydrate. This means that one molecule of $CaSO_4$ would be attached essentially to ½ molecule of water, hence the term *hemi*hydrate. *Hemi* refers to one half. When the reaction in the equation occurs, very small interlocking crystals of the dihydrate form and produce a slight increase in volume. This volume increase causes the new compound to tightly fit the mold into which it is poured and thereby form a sharp image of each coin or object.

Questions

1. What is a hydrate?
2. How many waters of hydration does plaster of Paris have?
3. Would you expect the appearance of hydrates to change if they were strongly heated? (Try some!)

Notes for the Teacher

BACKGROUND

Plaster of Paris, a common compound used in the building industry, is a form of calcium sulfate. When water is added to this compound, the water molecules are incorporated into the chemical structure of the compound to form a hydrate. Plaster of Paris is actually called a *hemihydrate* because each $CaSO_4$ is attached to ½ of a water molecule. When water is added and the dihydrate is formed, the compound increases slightly in volume and quickly hardens. We can use these properties to produce molds of some common objects such as coins.

TEACHING TIPS

1. Students may be interested in the origin of the name "plaster of Paris". "Plaster" comes from a Greek word meaning to form or to mold. Plaster of Paris was first made from calcium sulfate from Paris, France, hence the term "plaster of Paris".
2. Plaster of Paris is used extensively as plaster for the interior of buildings (it is mixed with small bits of animal hair to give it additional strength), for statues, and for wallboard.
3. Although other devices are now generally used, plaster of Paris casts have been used to set broken arms and legs. The material hardens quickly and molds to fit the contour of the arm or leg.
4. Blackboard chalk is usually calcium sulfate dihydrate. Actually, the term "chalk" is another word for calcium carbonate. The commonly occurring dihydrate of calcium sulfate, $CaSO_4 \cdot 2H_2O$, is also called "gypsum", which we named from the Greek word "gypsos", meaning "chalk". Therefore, we misuse the term "chalk" when referring to blackboard crayon and gypsum.
5. Students may be able to detect the increase in volume; however, it will only be a slight increase.
6. Encourage students to examine other hydrates. Hydrated copper sulfate (blue) can be heated to form the anhydrous form, which has a different physical appearance (white).
7. Try using cupcake cups, aluminum soft drink caps, and polystyrene cups as molds. Molds should be disposable and easy to tear away from the solid calcium sulfate. Students can also place the coins on top of the plaster of Paris.
8. You may want to mix the plaster of Paris in another paper cup to make cleanup easy.
9. Have students devise a way to make a mold of both sides of a coin.

ANSWERS TO THE QUESTIONS

1. A *hydrate* is a crystal that contains a salt and water combined in definite proportions.
2. Two; therefore, it is called a *dihydrate*.
3. Yes. Hydrated copper sulfate is deep blue, and the anhydrous form is white. See Activity 49 for a similar reaction using cobalt chloride.

14. Density and the Mystery Metals

Have you ever heard this proposition: "With which would you rather be hit in the face: a pound of feathers or a pound of lead?" Obviously, the answer is a pound of feathers! Although both weigh the same (1 lb), the feathers occupy a greater volume. A pound of feathers would be the size of a small pillow, whereas a pound of lead would be the size of a small (and very hard) ball. The density of the feathers is much less than that of the lead.

Density is this ratio of mass to volume. The units of density of solids often are grams per cubic centimeter (g/cm^3) or pounds per cubic foot (lb/ft^3).

In this activity you will determine the density of some common metals by measuring their mass and volume. The volume is determined by allowing the metal to displace water and then measuring the volume of water displaced. Finally, you can compare your calculated density to the chart to identify the mystery metal.

Materials

1. Metal samples: pieces of aluminum solid rod, iron (steel) screws and bolts, nails, fishing weights (lead), older solid silver or copper coins (pennies since 1982 are 97.6% zinc), items of jewelry, samples of metal elements from the chemical supply. Many small pieces of the same substance may act as one sample.
2. Balance.
3. Graduated cylinder, smallest available to hold the sample.

Procedure

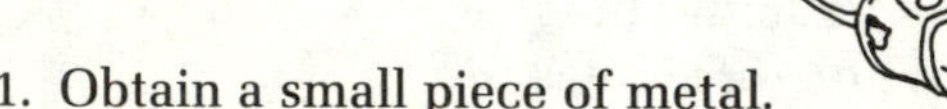

1. Obtain a small piece of metal.
2. Note any characteristic properties such as color and luster.
3. Carefully weigh the metal and record the mass.
4. Fill a graduated cylinder about one-half full with water. Read the volume and record your reading.
5. Place the metal in the cylinder. Do not allow any of the water to splash out of the container.
6. Notice that the metal has caused the water level to rise. Carefully record this new volume.
7. Subtract the smaller volume from the larger volume to get the volume of water displaced by the metal.
8. Use your data to calculate the density of your metal.
9. Compare your results to the chart to identify your metal.

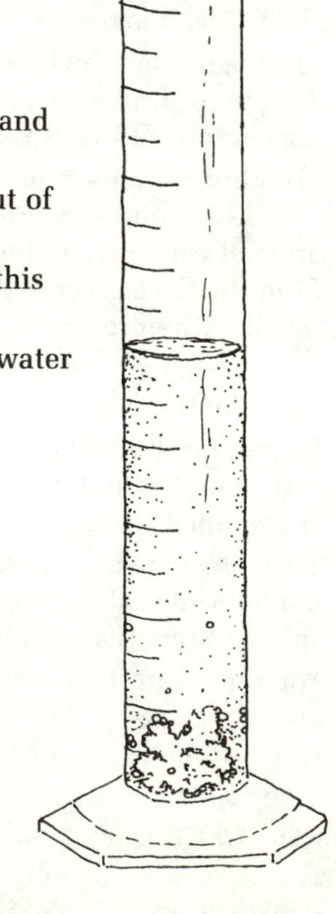

Calculations

Density is mass per unit volume:

$$\text{density} = \frac{\text{mass of metal (g)}}{\text{volume of water displaced (cm}^3\text{)}}$$

Questions

1. Could you determine the density of rocks with this same method? Try it!
2. Why can we use the volume of water displaced to determine the volume of the metal?
3. Which metal is the most dense? Which is the least dense?

Notes for the Teacher

BACKGROUND

The density of a substance is a characteristic physical property of that substance. Although the density may change slightly at a different temperature and pressure, this change is not significant. Density should not be confused with *specific gravity*, which is the ratio of the density of a substance to the density of a standard such as water or air. This activity gives students an introduction to determining density and using it to identify a substance.

TEACHING TIPS

1. Have students go to the library and read about Archimedes' famous bath tub experiment.
2. Have students bring in samples of rocks. Determine the density and arrange the rocks in order of density.
3. Prepare some unknowns (e.g., aluminum screws and lead fishing weights) for the students to examine.
4. See Activity 10 for another study of densities of liquids.
5. Densities (g/cm^3) of some common metals are as follows:

magnesium: 1.74	iron: 7.85
aluminum: 2.70	nickel: 8.90
chromium: 6.92	copper: 8.92
zinc: 7.00	lead: 11.35
manganese: 7.42	

6. Have students look up the densities of other metals such as gold and silver.

ANSWERS TO THE QUESTIONS

1. Yes, check the literature on densities of various rocks and minerals (e.g., quartz and sandstone).
2. The volume of water displaced is equal to the volume of the metal displacing it.
3. See Teaching Tip 5.

15. Tin Can Reaction

A "tin can" is actually an iron can with a very thin coating of tin. Tin is much less reactive with oxygen in the air than iron is; thus, it prevents the can from rusting and protects the contents of the "tin" can. Several methods are used to place a thin layer of one metal on another metal. In this activity we will examine one of these methods. You will not make an entire tin can; however, you will be able to place a coating of metallic tin on a piece of iron.

Materials

1. Iron nail.
2. Crucible.
3. Burner.
4. Metallic tin foil.
5. Zinc chloride ($ZnCl_2$).
6. Dilute nitric acid (HNO_3), 1 M.

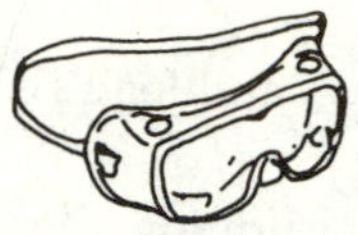

Procedure

1. Pack the crucible with small bits of tin foil until it is full.
2. In the hood, heat the crucible gently.
3. As the tin melts, carefully add more pieces until the crucible is about one-half full with molten tin.
4. Prepare an iron nail to be plated by dipping the nail into dilute nitric acid and then washing and drying. Avoid touching the acid.
5. Lower the temperature of the molten tin so that it is just above the melting point of tin.
6. Carefully sprinkle ½ tsp of zinc chloride on top of the melt. (If dense smoke is given off, the temperature is too high.)
7. Carefully dip the iron nail into the melt.
8. Rinse the nail several times with water. What do you observe? The bright coating is metallic tin.

Reactions

1. Melting occurs when the tin atoms gain enough heat energy to slide around each other. The melting point of tin is relatively low, 232 °C. The melting point of iron is 1535 °C.
2. The zinc chloride acts as a *flux*: It helps to remove any oxide from the surface of the iron nail prior to the tin plating.
3. The molten tin adheres to the iron, producing a thin and uniform coating.

Questions

1. Describe the tin coating on the iron nail.
2. What property of tin allows it to be plated on iron so easily?
3. How do you suppose larger iron objects are tin-plated?

Notes for the Teacher

BACKGROUND

In this activity students can coat an iron object (such as a nail) when a thin coating of metallic tin is placed on a thin strip of iron. The tin does not react as readily as iron does with oxygen in the air and thus protects the iron from rusting.

SOLUTIONS

Nitric acid is dilute. 1.0 M works well. See Appendix 2 for directions.

TEACHING TIPS

SAFETY NOTE: The hood is required for this activity because tin oxide fumes formed can irritate the nose.

1. If the temperature of the melt is too low, a glob of tin will plate on the nail. The molten tin may have an oxide coating, which should be skimmed off. Other features of the melt might be pointed out to students, for example, luster and surface tension.
2. The melting point of tin is 232 °C. The only common metal with a lower melting point is mercury, which is a liquid at room temperature.
3. Tin objects have been found in Egyptian tombs, and use of the metal dates back even to prehistoric times.
4. Tin is alloyed with copper to produce bronze, a metal that is much harder and more durable than either tin or copper.
5. The tin can was invented by Nicholas Appert, a scientist who lived during the time of Napoleon. Napoleon wanted a method of transporting food for his vast army over long distances without spoilage. Appert solved the problem by heating meat and wrapping it in tin-coated iron: the "tin" can!
6. The most important tin ore is cassiterite, SnO_2. Very little of this ore is found in the United States; however, the United States uses more than one-half of the tin produced in the world.
7. The tin coating on a tin can is one-fiftieth of the thickness of this page; however, this thickness is adequate to protect the iron inside from corroding.
8. Molten glass is floated on molten tin to make window glass.

ANSWERS TO THE QUESTIONS

1. The coating of tin will be uniform, thick, bright, and silvery.
2. Its low melting point (232 °C).
3. Strips of iron to be plated are dipped into containers of flux and then into large vats of molten tin.

Chemistry of Atoms and Molecules

16. Molecular Crystals: Sugar

In a solution, the molecules that are dissolved can be discovered by letting the liquid evaporate and by putting in seed crystals to which the dissolved molecules can cling. We call this "growing crystals". Many common chemicals can be used to grow large, gemlike crystals with definite geometric shapes. In this experiment, sugar crystals will be grown. **This activity will take 3–10 days.**

Materials

1. Table sugar.
2. Candy thermometer.
3. Pan to heat water and sugar.
4. Tall jar, pencil, string, and paper clip.
5. Burner.

Procedure

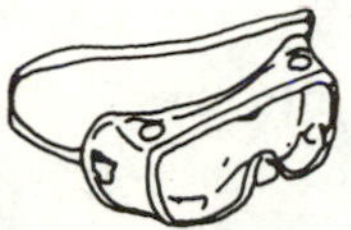

1. Heat 250 mL of water in a pan until it just begins to boil.
2. Record the temperature of the water every 30 s as it heats and for 2 min while it boils.
3. Gradually add 2 cups of sugar while continuing to take the temperature readings every 30 s. Do not stir with the thermometer.
4. Continue to heat and record the temperature until it reaches 115 °C (240 °F).
5. Carefully pour the hot solution into a tall jar.
6. Tie one end of a string to a pencil; tie the other end to a paper clip so that the paper clip almost touches the bottom of the jar when the pencil is placed across the top of the jar.
7. Dip the string into the hot solution to wet it.
8. Dip the string into dry sugar to cause a few "seed" crystals to attach to it.
9. Place the string in the jar and place the jar in a spot where it will not be disturbed.

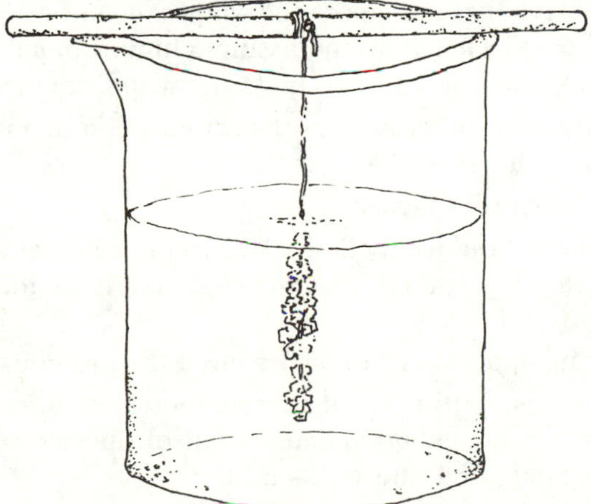

10. Make a graph of the temperature changes over time during heating and boiling by plotting the temperature on the vertical axis and the time on the horizontal axis. Mark the places on the line where (1) the pure water boiled, (2) the sugar was added, and (3) the sugar solution was boiling.
11. Look at the crystals each day (3–10 days). Record observations.
12. Remove the largest crystal from the string, measure it in millimeters, and make a drawing of it.

Reaction

Table sugar is called sucrose, $C_{12}H_{22}O_{11}$. When the crystals form, single molecules of sucrose come from the solution and attach to the surface of the sugar crystal to make the crystal become larger and larger. Growing crystals helps us to appreciate that substances are made up of single molecules. Solids have many molecules all piled up in regular arrangements in crystals. In solutions, the molecules that are dissolved are floating around among the water molecules.

Questions

1. What do you notice about the temperature of the pure water while it is boiling?
2. What do you notice about the temperature of the solution of sugar and water while it is boiling?
3. How large was your largest crystal?
4. Find out what other solids can be dissolved in water and then grown into crystals. Try some.

Notes for the Teacher

BACKGROUND

Many chemical principles are introduced in this activity, and you may want to emphasize some of them. The activity is also a lot of fun, and students experience the prolonged excitement of anticipation as they check the growth of the crystals each day. Basic principles involved are solution formation, supersaturation of a sugar solution, particle nature of substances, boiling, boiling point of a pure substance, boiling point of a solution, graphing, graph interpretation, and regularity of crystal shape.

TEACHING TIPS

1. Have students look at the sugar crystals and crystals of other solids with the binocular microscope. Have them compare the shapes. Be sure to look at salt (NaCl), potassium nitrate (KNO_3), alum (potassium aluminum sulfate or ammonium aluminum sulfate), and Rochelle salt (potassium sodium tartrate).
2. Comparison of the visible characteristics of the crystals can be made if the students carefully draw outlines of the crystals.
3. Do not allow students to eat this sugar.
4. Have students heat the solution evenly and take careful time–temperature data from the start of heating. They should have the sugar ready to add after the pure water has boiled for 2 min.
5. The jar, pencil, and string apparatus can be prepared the previous day.
6. Using a kitchen measuring cup (for the sugar) and metric volume (for the water) will allow the students to make general comparisons of the two volumes. They are similar (4 cups = 0.934 L, or 1 cup = 234 mL).
7. If students wish to repeat this activity outside the laboratory and if the string is replaced by a small wooden stick, a candy-coated stirring stick will be the result. These are popular at holiday time and are costly in the specialty shops.
8. Have students design other crystal-growing activities (see next activity). The salts mentioned previously are good ones to use. The seed crystal can be introduced by tying a perfect crystal to a piece of thread and suspending it in a jar of supersaturated solution.

9. Make comparisons of crystal growth rates of various substances, or vary the conditions for growing the same type of crystal.

ANSWERS TO THE QUESTIONS

1. Temperature is constant.
2. Temperature increases continuously.
3. Answer in millimeters.
4. See Activity 17.

17. Growing Crystals from Solutions

Many solids consist of atoms or molecules arranged in characteristic three-dimensional patterns. These patterns are repeated over and over to form certain geometric shapes that we call *crystals*. Crystals of a particular chemical substance usually have similar shapes, whether the crystals are as small as a pinhead or as large as an ice cube. In this activity you will make some small crystals by evaporating water from solutions to build up layers of atoms and molecules. **This activity will take 3 or 4 days.**

Materials

1. Sugar ($C_{12}H_{22}O_{11}$).
2. Potassium sulfate (K_2SO_4).
3. Ammonium ferric sulfate ($NH_4Fe(SO_4)_2 \cdot 12H_2O$).
4. Sodium ferrocyanide ($Na_4Fe(CN)_6 \cdot 10H_2O$).
5. Epsom salts [magnesium sulfate ($MgSO_4$)].
6. Baking soda [sodium bicarbonate ($NaHCO_3$)].
7. Eight large test tubes with one-hole stoppers to fit.
8. Eight circles of copper wire.
9. Thread.
10. Hot water.
11. Hand lens or magnifying glass.

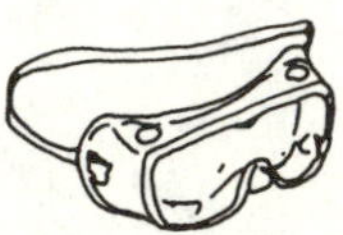

Procedure

1. Fill the first test tube half full with very hot water. Add sugar, a small amount at a time, and stir with a glass stirring rod until no more sugar dissolves.
2. Tie a 6-in. piece of thread to a circle of copper wire. Pass the other end of the thread through the hole of a one-hole stopper to fit the test tube.
3. Put the copper wire in the sugar solution and insert the stopper in the tube. Adjust the length of the thread so that the copper wire is about 0.5 in. from the bottom of the test tube. Hold the thread in place by taping it to the stopper with a small piece of masking tape. Keep the hole in the stopper open.
4. Set the tube aside where it will be undisturbed for 3 or 4 days.
5. Repeat the procedure with the following chemicals: potassium sulfate, ammonium ferric sulfate, magnesium sulfate (Epsom salts), sodium bicarbonate (baking soda), and sodium ferrocyanide. Add each chemical to a tube half full with hot water and stir until no more solid will dissolve. Be sure to label each tube.
6. After crystals have formed on each string, carefully remove each string from its tube and describe the crystals.
7. To help you describe the shape of each crystal, model patterns are provided. Cut out each model and fold it into the proper shape. These six geometric shapes represent some of the basic structures found in crystals.
8. Using these models as a guide, examine each crystal with a small hand lens or magnifying glass and describe completely each crystal, including its size, color, clarity, and shape. Compare your results with those of your classmates.

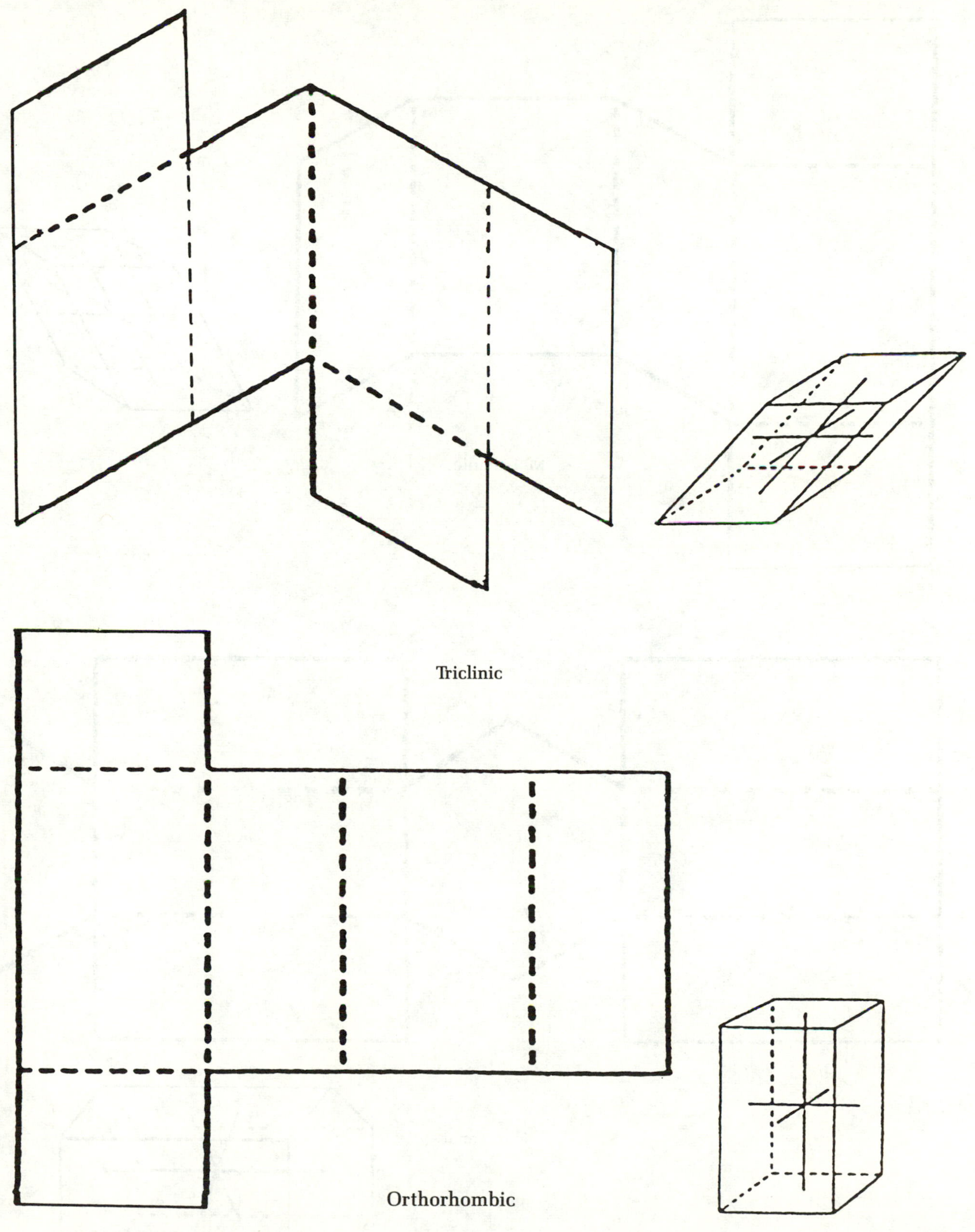

Triclinic

Orthorhombic

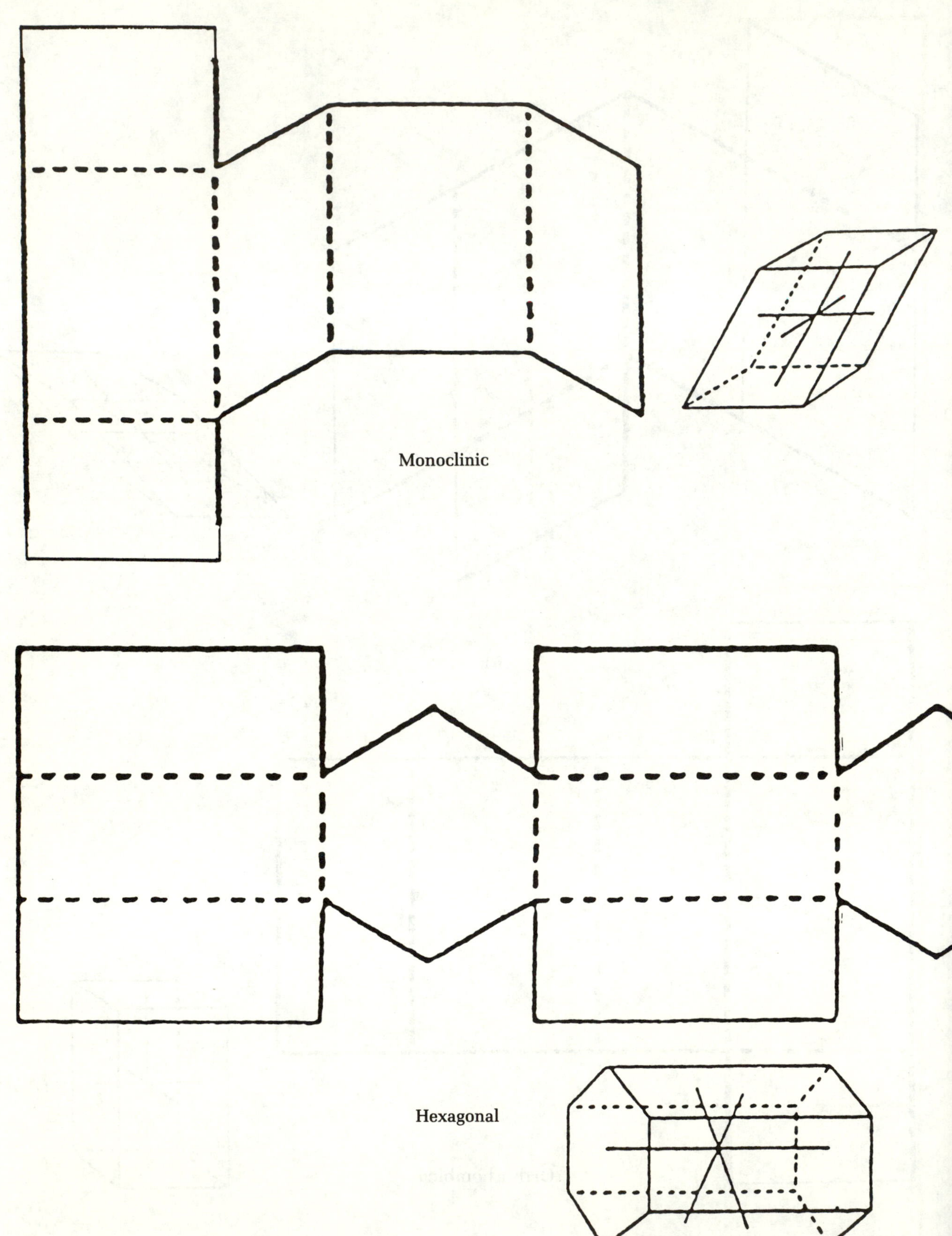

Monoclinic

Hexagonal

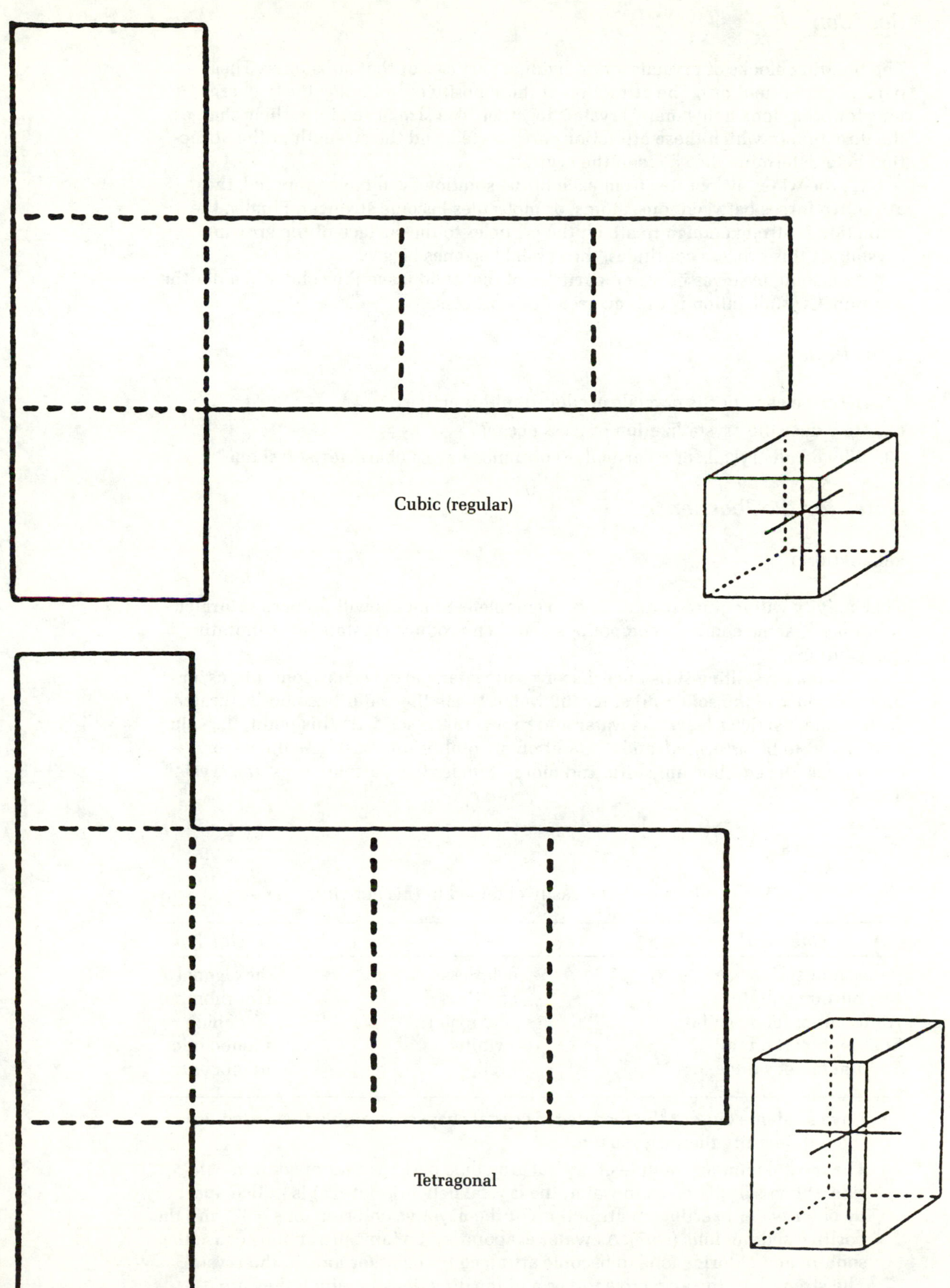
Cubic (regular)
Tetragonal

Reaction

The building blocks of crystals are individual particles of that substance. These particles are often *ions.* The attraction of these positively and negatively charged ions for other ions holds many crystals together. The size of the ions, their charges, the direction in which these attractions are exerted, and the strength of the attraction help determine the shape of the crystal.

As the water evaporates from each of the solutions you have prepared, the attractive forces between ions, atoms, or molecules become stronger. Finally, the attraction is strong enough to attach the particles to the surface of the growing crystal. As this process continues, the crystal becomes bigger.

As a solid dissolves in water, particles of that solid leave the solid and enter the solution. Crystallization is the reverse of this process.

Questions

1. Describe each of the crystals produced in this activity.
2. How does the crystallization process occur?
3. Why do all crystals of a particular substance have a characteristic shape?

Notes for the Teacher

BACKGROUND

This activity will require several days to complete. Students will prepare saturated solutions of some common compounds and then produce crystals by evaporating the solutions.

When a crystalline substance dissolves in water, particles (i.e., ions, atoms, or molecules) leave the solid and enter the water. When the water becomes saturated with solute, particles leave the water and return to the solid. At this point, the solution is said to be *saturated*, and a condition of equilibrium exists. As the water evaporates, the equilibrium shifts and more particles form a solid; thus, the crystal grows.

TEACHING TIPS

1. The following results should be easily obtained in this activity:

Chemical	*Color*	*Crystal Type*
Potassium sulfate	colorless	hexagonal
Magnesium sulfate	colorless	orthorhombic
Ammonium ferric sulfate	violet (pale)	cubic
Sodium bicarbonate	white	monoclinic
Sodium ferrocyanide	yellow	monoclinic

2. Have students cut out the models for crystal shapes, assemble them, and use them to identify their crystals.
3. The most common example of crystal growth is perhaps that of sodium chloride. When salt dissolves in water, the crystal network (lattice) is pulled apart by polar water exerting an attraction for the negative chloride ions (Cl^-) and the positive sodium ions (Na^+). As water evaporates, the ionic attractions cause the sodium and chloride ions to become attracted to the other ions in the crystal. The strength of the attractive forces and the directions in which they are exerted cause the formation of the cubic crystal.

4. Some other crystals and their characteristic forms are as follows:

CUBIC:

sodium chloride ($NaCl$)
potassium chloride (KCl)
sodium phosphate ($Na_3PO_4 \cdot 10H_2O$)
*sodium chlorate ($NaClO_3$)
*sodium bromate ($NaBrO_3$)
ammonium chloride (NH_4Cl)

ORTHORHOMBIC:

*potassium nitrate (KNO_3)
zinc sulfate ($ZnSO_4$)
*potassium permanganate ($KMnO_4$)
potassium sodium tartrate ($KNaC_4H_4O_6$

MONOCLINIC:

sodium sulfate (Na_2SO_4)
sodium carbonate (Na_2CO_3)

TRICLINIC:

copper sulfate ($CuSO_4 \cdot 5H_2O$)
calcium hexacyanoferrate(II) [$Ca_2Fe(CN)_6$]
*calcium nitrate [$Ca(NO_3)_2 \cdot 3H_2O$]

TETRAGONAL:

*zinc nitrate [$Zn(NO_3)_2 \cdot 6H_2O$]
ammonium copper chloride ($NH_4Cl \cdot CuCl_2 \cdot 2H_2O$)

HEXAGONAL:

zinc iodide (ZnI_2)
calcium iodide ($CaI_2 \cdot 6H_2O$)

*Strong oxidizing agents.

5. A seventh crystal system, rhombohedral, is exhibited by $CaCO_3$. Actually, 14 crystal lattices are possible, but they are all variations of the seven basic forms.
6. Try dissolving together potassium sulfate and sodium sulfate. The crystals that result are $(K_2SO_4)_3 \cdot Na_2SO_4$, which are rhombohedral.

ANSWERS TO THE QUESTIONS

1. See Teaching Tip 1.
2. As water evaporates, attractive forces are exerted between solid particles, causing them to become attracted and attached. In most cases the forces are ionic forces; however, in some molecular solids such as sugar, the forces are weak intermolecular forces.
3. The attraction of positive and negative particles for each other causes a particular pattern to form. This situation produces a regular growth in a particular direction and a shape of the crystal.

18. Delicate Crystal Growth in Gels

When a gel is made from sodium silicate, it is a semisolid polymer. Sodium silicate is also called *water glass* because it forms a glasslike coating over eggs to preserve them by sealing the pores in the shell. In this activity you will grow crystals of copper in the gel through a chemical reaction. The crystals are very delicate and are held in place by the gel molecules in which they have grown. You will have a chance to compare the growth rate of copper crystals when different metals are placed in the gel tubes made with copper sulfate solution. **This activity will take 2 days.**

Materials

1. Diluted sodium silicate (Na_2SiO_3) solution.
2. Several medium-size test tubes with rubber stoppers.
3. Acetic acid (HCH_3COO) solution, 1.0 M.
4. Copper sulfate ($CuSO_4$) solution, 1.0 M.
5. Sodium chloride (NaCl, table salt) solution, 10%.
6. Pieces of metal: iron (Fe) nail, piece of tin (Sn) or tin foil, piece of lead (Pb), piece of zinc (Zn), piece of magnesium (Mg) ribbon, and piece of aluminum (Al) cut from a pie pan. Clean each metal with steel wool to remove the oxide layer.
7. Stirring rod.

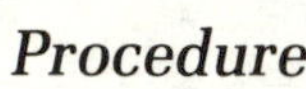

Procedure

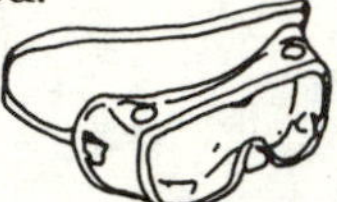

1. Prepare a test tube for each metal to be investigated by placing 10 mL of acetic acid solution in the tube.
2. Place 25 drops of copper sulfate solution in each tube.
3. While stirring the solution, add 10 mL of sodium silicate solution to each test tube. Avoid spilling the solution; sodium silicate is corrosive.
4. Allow the tubes to set overnight or place them in a warm water bath to shorten the setting time.
5. When the gel forms, drop in a piece of metal and push it under the surface of the gel.
6. Place 5 drops of sodium chloride solution in the tube and put in a rubber stopper.
7. Observe your tubes for several days. Record observations and changes in a data table.

Reaction

When in contact with these metals, copper atoms are most stable if they have all of their electrons. The copper atoms in blue copper sulfate solution are missing two electrons each. In this state the atoms are called *ions*. Each metal placed in the gel will give electrons to the copper ions, and the metal atoms that give up their electrons will enter the solution as ions of that metal.

Example:

$$Zn(s) + Cu^{2+}(aq) \longrightarrow Cu(s) + Zn^{2+}(aq)$$

Questions

1. List the changes that you see when copper ions are changed to copper atoms.
2. In which tube does the copper form the fastest?
3. Which metal do you think is the most active in giving electrons to copper?
4. Write reactions, such as the one given in the Reaction section, in which you substitute other metals for zinc.

Notes for the Teacher

BACKGROUND

Gels make interesting media for growing crystals. Because diffusion is somewhat slow in gels, the growth of the crystals takes place over a long time, although growth is apparent from the beginning. The reactions in the tubes are oxidation–reduction reactions. In an oxidation–reduction reaction, one kind of atom gives electrons to another. You could also do precipitation reactions in which two kinds of ions bond to form an insoluble solid crystal.

SOLUTIONS

1. For sodium silicate, use commercial water glass obtained from the pharmacy or hardware store.
2. Dilute the sodium silicate to a density of 1.06 g/cm^3 (about 1 part of sodium silicate to 4 parts of distilled water). This solution is corrosive.
3. Acetic acid: Dissolve 30 mL of concentrated acetic acid in water and dilute to 500 mL.
4. Copper sulfate: Dissolve 125 g of copper sulfate pentahydrate ($CuSO_4 \cdot 5H_2O$) in water and dilute to 500 mL.
5. Sodium chloride: Dissolve 10 g of sodium chloride in water and dilute to 100 mL.

TEACHING TIPS

1. Test tubes should hold about 25 mL.
2. The sodium chloride solution helps to provide a flow of electrons.
3. Using uniform iron nails, you might have students experiment with various concentrations of copper sulfate to find the relationship between concentration and rate of reaction or amount of product.
4. Students might experiment with some tubes in which two different metals are placed in copper sulfate gel.

ANSWERS TO THE QUESTIONS

1. The color changes, and the solid forms in a crystal arrangement.
2. In the tube with magnesium.
3. Magnesium.

4. The reactions are as follows:

$$2Fe(s) + 3Cu^{2+}(aq) \longrightarrow 2Fe^{3+}(aq) + 3Cu(s)$$

$$Sn(s) + Cu^{2+}(aq) \longrightarrow Sn^{2+}(aq) + Cu(s)$$

$$Pb(s) + Cu^{2+}(aq) \longrightarrow Pb^{2+}(aq) + Cu(s)$$

$$Mg(s) + Cu^{2+}(aq) \longrightarrow Mg^{2+}(aq) + Cu(s)$$

$$2Al(s) + 3Cu^{2+}(aq) \longrightarrow 2Al^{3+}(aq) + 3Cu(s)$$

19. Salt Crystal Garden

When salts are dissolved in water, they cannot be seen. However, if the water molecules are allowed to evaporate from the solution, the salt crystals will form small particles that are visible. Table salt crystals often form perfect cubes that can be identified if viewed under a microscope. Seawater can be evaporated to form a mixture of various crystals. When table salt is mixed with *bluing*, a product used in rinsing white fabrics to prevent yellowing, some very interesting crystals form. See what kinds of formations develop when you follow this procedure. **This activity will take 3–5 days.**

Materials

1. Something porous: a piece of brick, lump of charcoal, or piece of cement.
2. Table salt, 2 Tbsp.
3. Water, 4 Tbsp.
4. Laundry bluing, 2 Tbsp.
5. Household ammonia, 2 Tbsp.
6. Microscope or hand lens.

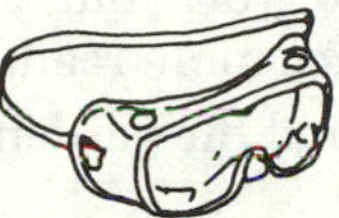

Procedure

1. Put the piece of brick or other porous material into a shallow bowl.
2. Mix the four chemicals. Pour them over the brick.
3. Three days later, mix the same amount of the four chemicals and add to the bottom of the bowl.
4. Compare the crystals that form with the crystals of pure table salt by viewing both under the microscope or with a hand lens.
5. See how long you can continue the "garden" by adding the same amount of the chemicals periodically.
6. If you wish, you may add food coloring to the top of the brick.

Reaction

All of the changes that you see are physical changes of salts dissolving and recrystallizing. The bluing is a suspension of particles of a blue powder called iron(II) hexacyanoferrate(III). The tiny particles of the blue powder in the water act as "seed" particles around which the salt crystals form. When table salt forms around these particles, it forms starlike crystals rather than cubes. The crystals are very fine. Water molecules and the dissolved salts move through the brick because they tend to cling to each other. This movement is called *capillary action*.

Questions

1. Draw the shapes of the pure salt crystals from the box of salt and the shapes of the garden crystals.
2. How could a sample of solid sea salt crystals be obtained?
3. What process do you think is being described when one says that salt crystals grow?

Notes for the Teacher

BACKGROUND

This activity is a good way to generate discussion about the existence of particles too tiny to be seen with the unaided eye. Students should be given the opportunity to imagine the presence of invisible things for many reasons, one of which is the importance of accepting abstract ideas in chemistry in order to understand other aspects of the field. Another reason for students to acknowledge the existence of particles on the atomic, molecular, and ionic levels is so that they can understand issues such as pollution of the environment. In this simple activity, which has been popular since the childhood days of the students' grandparents, the beauty of common crystals is fun to observe.

TEACHING TIPS

1. The ammonia is added to increase the solubility of the salt. It can be omitted.
2. Some air circulation is necessary. Winter is a nice time to grow your garden because the air is dry and thus the evaporation takes place more readily.
3. Continue to add the mixture to the bottom of the bowl, and the garden will continue to form.
4. The product, bluing, is also used as an ink to mark clothing and to mark logs in the timber industry.
5. You might make a small amount of concentrated solution of table salt and evaporate it. Large crystals will form and are fun to look at under the binocular scope or any microscope set at low power.
6. A few years ago salt manufacturers put a very small amount of the iron(II) hexacyanoferrate(III) fine powder in boxes of table salt to cause fine crystals of salt to form instead of the clumped ones that seem to form during humid weather. The formula for this chemical is $KFe_2(CN)_6$.
7. Activity 20 may be done at the same time.

ANSWERS TO THE QUESTIONS

1. Pure salt often forms a perfect cube. Garden salt is star-shaped.
2. The sample would be a mixture of crystals, but it could be obtained by evaporating seawater. Seawater is a 3% mixture of various salts, most of which is sodium chloride.
3. The dissolved particles are arranging themselves onto the seed crystal in a regular repeating pattern called *crystallization*.

20. The Chemistree: Absorption and Evaporation

In this activity you will prepare a solution and allow it to be absorbed through blotter paper. As the liquid part of the solution evaporates, the solid will remain in the form of beautiful crystals—the chemistree! This activity takes 1–2 days.

Materials

1. Blotting paper (green, preferably) or lightweight cardboard.
2. Petri dish, one-half dish per student, or shallow bowl.
3. Laundry bluing.
4. Table salt.
5. Ammonia.
6. Food coloring.

Procedure

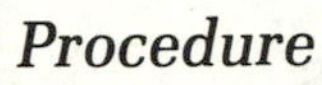

1. Cut three pieces of blotter paper in the shape of a fir tree. Cut a small slit at the bottom of two pieces and a slit in the top of one piece as shown in the illustration. Each piece should be about 4 in. high. The base should be just small enough to fit into a small shallow dish.

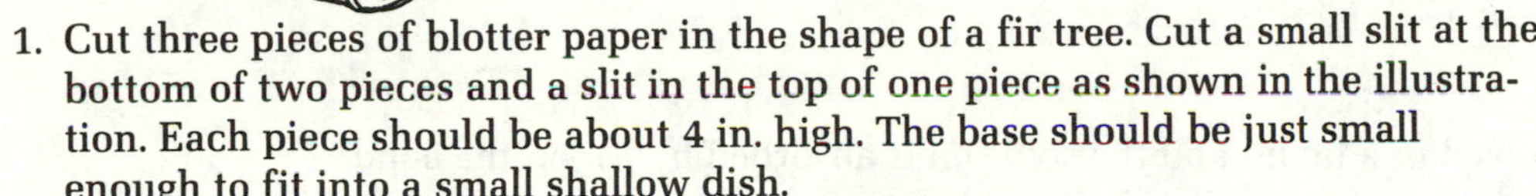

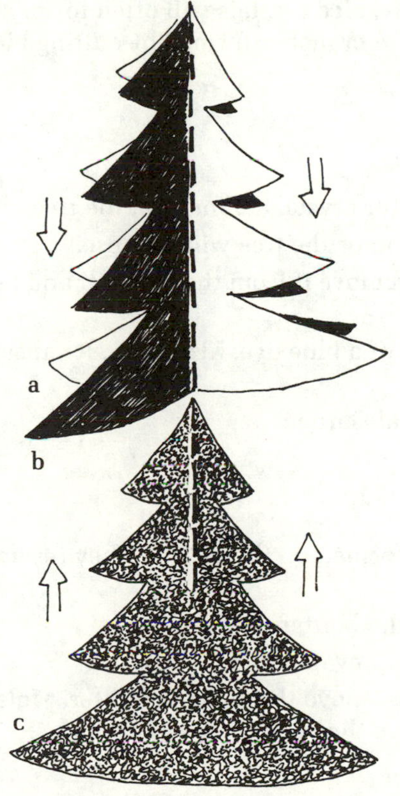

2. Insert pieces a and b on top of piece c to make a tree.
3. Prepare a special solution containing the following ingredients: water (6 Tbsp), laundry bluing (6 Tbsp), ammonia (1 Tbsp), and table salt (6 Tbsp).
4. Fill the shallow dish with the special solution.
5. Place the base of the tree in the dish so that the tree stands upright.
6. Place 1 drop of food coloring at the tip of each branch of the tree.

7. Observe the tree for several hours.
8. Explain your observations.

Reaction

Crystals of salt dissolve in the solution. These dissolved crystals are absorbed with the solution onto the blotter paper and travel up the blotter. Evaporation occurs faster at the edge of the blotter and results in crystal formation.

Questions

1. What did you observe?
2. How did the solid get to the branches of the tree?
3. How did the crystals form?
4. Why did it take so long for this reaction to occur?

Notes for the Teacher

BACKGROUND

When a solid is dissolved in a liquid and the solution is absorbed by paper, the solid travels up and through the paper. As the liquid (solvent) evaporates, the solid is left behind. If this process is done slowly, nice crystals will often form. In this activity, crystals will appear on the tips of the branches formed by cutting blotting paper into fir tree shapes.

TEACHING TIPS

1. Several hours may be required for crystals to form on the tree.
2. Do not let too many crystals form or the tree will collapse.
3. When the tree is fully formed, remove it from the solution and keep it as a permanent display.
4. Laundry bluing is a suspension of a blue dye, $KFe_2(CN)_6$. It causes the salt crystals to be very fine.
5. See also Activity 19, "Salt Crystal Garden".

ANSWERS TO THE QUESTIONS

1. After a few hours, the crystals formed on the blotter. They resembled snow on the branches of a fir tree.
2. It was absorbed in solution by the blotter.
3. Crystals formed when the solvent evaporated.
4. Molecules of water and salt must move through the blotter. Molecules of water must gain enough energy to leave the solution (evaporation).

21. Sulfur Makes a Nice Impression

Sulfur is an interesting element. It exists in several different forms. Different forms of the same substance are called *allotropes*. We will examine two of these allotropes in this activity. Ordinary yellow sulfur (native sulfur) is called *rhombic sulfur*. When this sulfur is melted and allowed to crystallize, it forms another type of sulfur called *monoclinic sulfur*. Monoclinic sulfur has long needles. We will melt sulfur and pour the liquid sulfur into a mold containing a coin. When the sulfur cools, we will have a nice impression of the coin!

Materials

1. Penny or other coin.
2. Sulfur.
3. Paper cups (unglazed Dixie Cups work well).
4. Burner.
5. Art blade or scissors.
6. Tongs.

Procedure

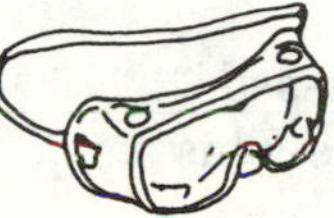

1. First, prepare the mold for your experiment. On the bottom of a small paper cup, outline a penny or other coin by drawing around its outer edge with a pencil or pen. Cut out the circle to form a window and hold the cup firmly on top of the coin. The coin should have a tight fit at the bottom of the cup.
2. Fill the test tube about three-fourths full with sulfur. Note the properties of this sulfur.
3. Slowly heat the sulfur just until it melts. Do not overheat.
4. When the sulfur has completely melted, carefully and quickly pour it into the paper cup while holding the paper cup with tongs tightly in place over the coin.
5. When the sulfur has cooled and hardened, carefully tear away the paper cup. Did the coin leave a nice impression in the sulfur? How does the crystallized sulfur compare with the original sulfur?

Reaction

This reaction is merely a phase change from one allotropic form of sulfur to another. Ordinary crystals of yellow sulfur are orthorhombic in shape.

At 95.2 °C, sulfur melts. The straw-colored liquid is called *lambda sulfur*. When it cools and crystallizes, this sulfur forms monoclinic sulfur. Crystals of this form of sulfur are long needles.

Eventually, all forms of sulfur at room temperature change back into the ordinary yellow rhombic form.

Refer to Activity 17 for more on these shapes.

Questions

1. What is an allotrope?
2. What allotropes of sulfur did you examine?
3. What property of sulfur permitted you to make an impression of a coin?
4. Do you know of other elements that have other forms (allotropes)? Go to the library!
5. What is the melting point of sulfur?

Notes for the Teacher

BACKGROUND

Some elements exist in different forms. These different forms are called *allotropes*. Sulfur has several allotropes. In the normal, yellow form it is called *rhombic sulfur*. When heated, it melts to form *lambda sulfur*. When this form is allowed to cool, it crystallizes into *monoclinic sulfur*. If the molten lambda sulfur is poured into water to cool it rapidly, it forms *plastic sulfur*.

TEACHING TIPS

1. Have students devise other methods to prepare a mold for the coin.
2. Sulfur at room temperature actually consists of eight atoms of sulfur connected in a puckered ring, S_8. Even though chemists know this fact, they usually represent sulfur simply as "S" in chemical equations.
3. Sulfur occurs in the free form, and large deposits are found in Texas and Louisiana.
4. Sulfur is extracted by forcing superheated water into the ground. Compressed air is then forced into the melt, and molten sulfur is pushed to the surface. This process is called the *Frasch process*.
5. Students should not overheat the sulfur. When melted, it will be pale yellow. If students continue to heat the liquid, it will become darker and syrupy at 150 °C.
6. Sulfur is the brimstone described in the Bible.

ANSWERS TO THE QUESTIONS

1. An *allotrope* is a different form of the same element.
2. In this activity, the rhombic and monoclinic forms were examined.
3. The quick cooling of the sulfur and the fine crystal formation caused an impression of the coin to form.
4. Yes. Arsenic, carbon, oxygen, phosphorus, and tin also form allotropes. (Encourage students to look these up and report on them.)
5. The melting point of rhombic sulfur is 95.2 °C.

22. Reaction of Copper and Sulfur

When orange-red copper and yellow sulfur react, several surprising changes take place. See how many changes you can observe. You will also check to see whether or not a change in mass occurs when a new substance forms.

Materials

1. Glass test tube, 18 × 150 mm. (Don't use soft glass.)
2. Powdered sulfur (S), 1 g.
3. Granular copper (Cu), 2 g.
4. Small balloon.
5. Burner.
6. Balance.
7. Dilute hydrochloric acid, 0.1 M.

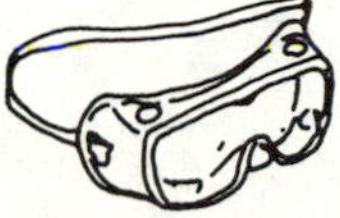

Procedure

1. Examine the copper granules and sulfur powder with a hand lens.
2. Mix the copper and sulfur and place them in the test tube.
3. Fix a balloon over the mouth of the test tube, using a rubber band, if necessary, to hold the balloon on the test tube.
4. Find the mass of the tube with its contents and balloon.
5. Heat the mixture gently until it begins to glow.

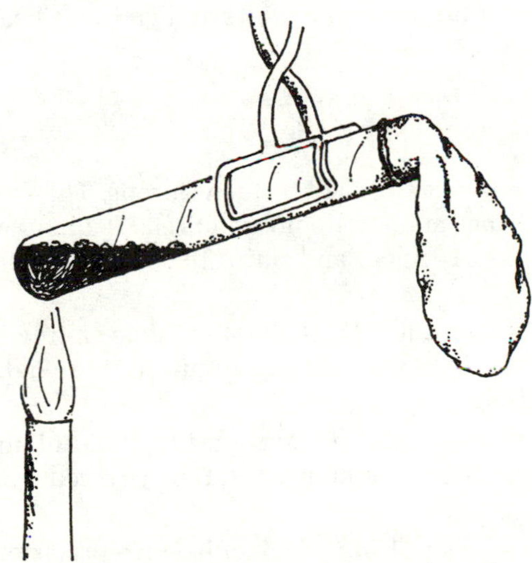

6. Remove the burner immediately and watch the tube glow on its own.
7. Allow the tube to cool.
8. Again find the mass of the tube with its contents and balloon.
9. Examine the contents of the tube.
10. Compare the new substance with some copper and with some sulfur. Add a little dilute hydrochloric acid to a sample of each of the three substances. Describe what happens.

Reaction

Copper reacts with sulfur to make black copper sulfide. The reaction gives off energy as heat and light.

$$\underset{\text{copper}}{Cu(s)} + \underset{\text{sulfur}}{S(s)} \longrightarrow \underset{\text{copper sulfide}}{CuS(s)} + \text{heat and light}$$

Questions

1. What makes you think the black product is a new substance?
2. What does hydrochloric acid do to copper sulfide? What does it do to copper?
3. Did all the copper and sulfur react? How do you know?
4. Compare the mass of the reactants (copper and sulfur) with the mass of the products (copper sulfide) in this reaction.
5. What is meant by conservation of mass?

Notes for the Teacher

BACKGROUND

Copper metal is not as reactive as many other metals; however, it will react with vigor with sulfur if the reaction is started by heat from a burner. This simple reaction introduces the common elements, copper and sulfur, and the ideas of chemical change, law of conservation of mass, and exothermic reactions. We know the reaction is *exothermic* (i.e., energy-releasing) because it glows and gives off heat after the burner is removed.

TEACHING TIPS

1. Do not use powdered copper. The reaction will be too vigorous. The copper and sulfur need to be mixed well. Exact masses are not required. You may want to have students weigh out samples of copper and sulfur the day before this activity.
2. Watch carefully for the glow to occur. It is brief, and the flame should be removed right away. It is fun, if safe in your setting, to turn out the lights when everyone is ready to begin heating.
3. The balloon traps the gases. Students will find it interesting that hot air may fill the balloon. Someone may want to heat an empty test tube covered with a balloon to see.
4. Any apparent mass changes may result from a lack of balance precision.
5. This reaction of copper and sulfur is the opposite of the reaction that produces copper from the copper ore, copper sulfide.
6. Some sulfur dioxide is also formed when the test tube is heated, but this gas is contained in the balloon. In the environment, SO_2 is produced when sulfur in coal is burned in power plants. SO_2 combines with oxygen and rainwater to form aqueous sulfuric acid, or rather, "acid rain".

$$SO_2(g) + \tfrac{1}{2}O_2(g) \longrightarrow SO_3(g)$$

$$SO_3(g) + H_2O(\ell) \longrightarrow H_2SO_4(aq)$$

ANSWERS TO THE QUESTIONS

1. The new substance is a new color (black).
2. Copper sulfide reacts with hydrochloric acid to form a green solution, copper chloride. Copper and sulfur will not react with hydrochloric acid.
3. Not all of the copper and sulfur reacted. When the acid was added, the copper sulfide was removed from the surface of the copper granules, and the copper was exposed.
4. Mass should not change.
5. Mass is not created or destroyed in chemical reactions.

23. Putting Atoms Together: Synthesis of Zinc Iodide

In a chemical reaction, atoms often combine to form compounds. Chemical properties of the compounds are quite different from those of the atoms of which they are composed. In this activity you will take two elements, zinc and iodine, and combine them to form an ionic compound, zinc iodide. When atoms combine, energy is often involved in the formation of compounds. We can lower the amount of energy required by using a catalyst. In this activity, water is the catalyst that initiates the reaction.

Materials

1. Powdered zinc (Zn).
2. Iodine crystals (I_2).
3. Funnel and filter paper.
4. Watch glasses or dishes.
5. Dropper and small beaker.

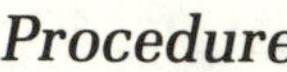

Procedure

1. Examine the powdered zinc and iodine crystals. Record their properties.
2. Weigh out about 0.2 g of zinc powder. Place the powder in the watch glass.
3. Weigh out the same amount of iodine crystals. Place the crystals in the watch glass also. Mix the two by stirring with a glass rod or wood splint. Is any evidence of a reaction present?
4. Place the watch glass in a vented hood on a piece of white paper.
5. Carefully add ½ dropper full of water to the mixture. Is evidence of a chemical reaction present? CAUTION: Iodine vapors are toxic.
6. When all of the water has been added, stir the solution with a glass stirring rod. Describe the material.
7. Filter the solution, being careful to transfer all of the product to the filter paper. Catch the filtrate in a clean beaker.

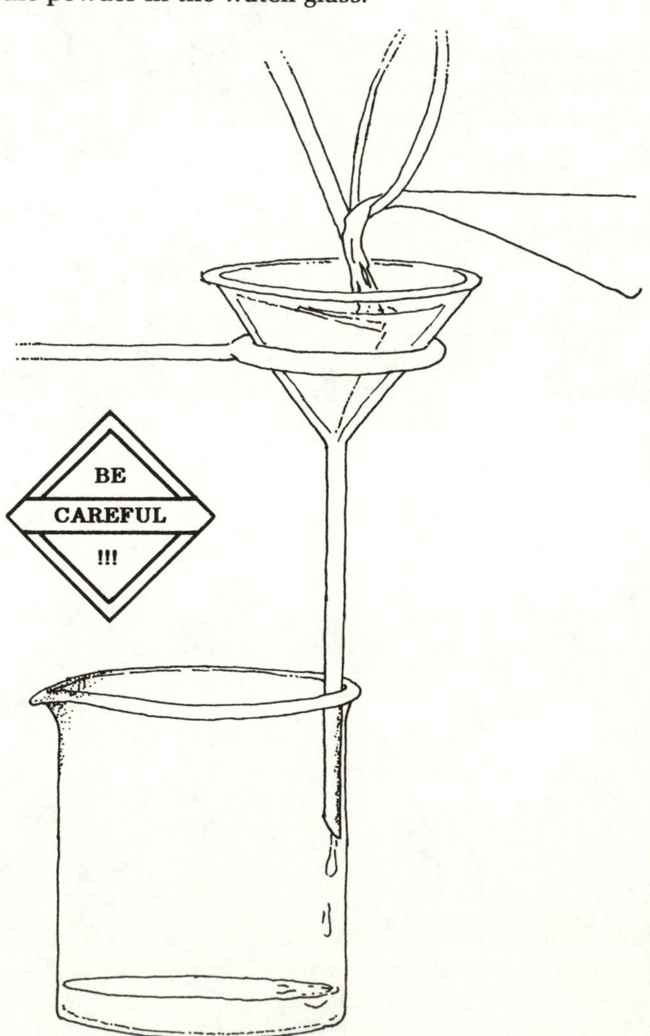

8. Discard the material on the filter paper and carefully pour the filtrate into a clean watch glass.
9. Place the watch glass in an open area where the water can evaporate. You can heat the dish with a heat lamp to increase the rate of evaporation.
10. When the liquid has evaporated, examine the product remaining in the dish. Describe this product. What is the name of this product? How does it differ from the reactants?
11. Do not discard the product. You can use it in the next activity to re-form zinc and iodine.

Reactions

A small amount of energy is required to start the reaction between zinc and iodine. Water acts as a catalyst to permit the reaction to begin at room temperature. When the reaction begins, it provides enough energy to melt and vaporize the iodine:

$$Zn(s) + I_2(s) \xrightarrow{H_2O} ZnI_2(s)$$

Some of the iodine is vaporized by the reaction. This vaporization accounts for the violet "smoke".

Questions

1. What evidence of a chemical reaction did you observe when water was added to the mixture?
2. Describe the compound that resulted when the filtrate was evaporated.
3. Write the chemical reaction that occurred to produce this new product.

Notes for the Teacher

BACKGROUND

This activity allows the student to produce a compound by a simple reaction between iodine and zinc. When water is added to a mixture of the two, zinc iodide is produced. This compound is a white solid that can be separated from the mixture by filtering and then evaporating the filtrate.

The product of this reaction, zinc iodide (ZnI_2), consists of zinc and iodide ions. Therefore, strictly speaking, this product is not composed of molecules because molecules consist of atoms. Nevertheless, we will call it a compound (an ionic compound is a more precise description).

When this compound is formed, the zinc atom loses two electrons. Thus, the zinc atom is oxidized.

$$Zn(s) \longrightarrow Zn^{2+} + 2e^-$$

Two iodine atoms pick up one electron each from the zinc atom. Thus, the iodine atoms are reduced.

$$I_2(s) + 2e^- \longrightarrow 2I^-$$

Therefore, this reaction is actually an oxidation-reduction reaction. We represent the ionic compound formed simply as ZnI_2.

MATERIALS

1. Zinc powder: The smaller mesh works best.
2. Iodine crystals: If the crystals are large, grind them into a powder with a mortar and pestle.

TEACHING TIPS

SAFETY NOTE: Iodine vapors will be produced. This reaction must be done in a hood or in a well-ventilated area.

1. When the water is added to the mixture of zinc and iodine, most of the material will remain unreacted as a gray slurry. Some spattering may occur.
2. This activity gives students experience with the simple techniques of filtering and evaporating.
3. You may or may not want to introduce oxidation–reduction reactions at this point. You may want to emphasize only the formation of a compound from its elements.
4. Be careful when evaporating the filtrate. Overheating will cause spattering and loss of the white product.
5. If students place the filtrate in a white porcelain evaporating dish, they may not be able to see the white solid that is produced. Therefore, we suggest that you use a clear watch glass or beaker.
6. Save the white solid, zinc iodide, for the next activity in which it can be decomposed into the original elements, zinc and iodine.
7. When iodine forms an ion, we call the ion *iodide*, not iodine. Thus, the name of the product is zinc iodide.

ANSWERS TO THE QUESTIONS

1. Heat was produced, which caused violet iodine vapors to form; the addition of water caused some fizzing.
2. The zinc iodide that remained when the filtrate was evaporated was a white solid, unlike either the gray zinc or the violet iodine.
3. See the Reactions section.

24. Taking Compounds Apart: Decomposing Zinc Iodide

In Activity 23, you made the compound zinc iodide by reacting metallic zinc and iodine crystals. In this activity you will decompose the zinc iodide to produce the original zinc and iodine. Because energy is required to break this compound into its constituent parts, we will use a battery. This process is called *electrolysis* and is often used to separate a compound into its components.

Materials

1. Sample of zinc iodide (ZnI_2) prepared in Activity 23.
2. Copper wire.
3. Battery. Any 6–12-V battery works well.
4. Battery holder and wire leads to connect the copper wires to the battery terminals.
5. Masking tape or modeling clay.
6. Glass stirring rod.

Procedure

1. Prepare a sample of zinc iodide according to the directions in Activity 23.
2. Using a dropper or a small pipet, add a small amount of water to the watch glass containing the white precipitate. Carefully stir with a clean glass rod until all of the solid dissolves.
3. Place two small pieces of copper wire in the solution in the watch glass. The two wires should be about 1 in. apart. Attach them to the watch glass with a small amount of masking tape or clay to prevent them from moving.
4. Attach the other ends of the two copper wires to the terminals of a battery, using the wire leads.

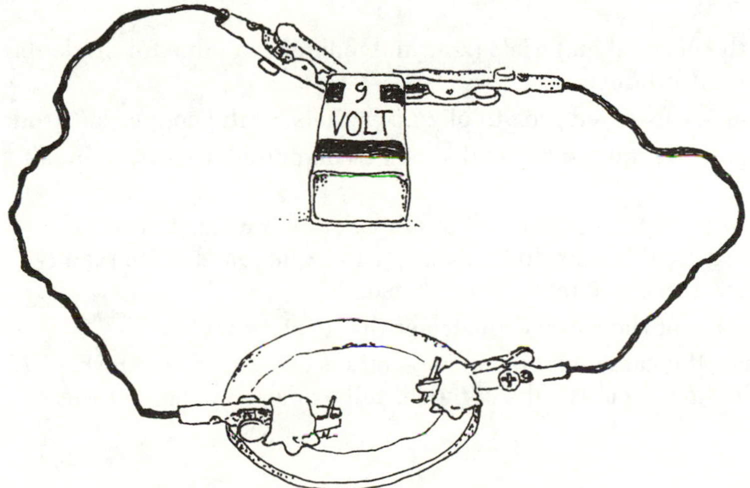

5. Observe. Record your observations.

Reactions

Two reactions are occurring. One reaction occurs at the *cathode*. Electrons from the battery flow into the solution at the cathode. This process produces metallic zinc, which accumulates on the copper cathode as a zinc coating:

$$Zn^{2+}(aq) + 2e^- \longrightarrow Zn(s)$$

The other electrode is the *anode*. Electrons return from this electrode to the battery. At this electrode, the iodide ions lose electrons and form iodine.

Iodine crystals will be seen forming on the anode, and they will dissolve in water as well:

$$2I^- \longrightarrow I_2(s) + 2e^-$$

The iodine causes a water solution to turn amber.

Questions

1. What evidence of a chemical reaction is present?
2. How do you know that you have produced zinc and iodine?
3. What role did the battery play in this reaction?

Notes for the Teacher

BACKGROUND

This activity is a simple electrolysis experiment that produces zinc and iodine from a solution of zinc iodide. It is most effective when coupled with Activity 23, in which zinc iodide is synthesized.

TEACHING TIPS

1. Have samples of the original materials (zinc and iodine) available for students to compare with their products.
2. Zinc will appear as a whiskered growth of zinc crystals on the copper electrode.
3. Point out that energy is required to break down a compound, that is, to break chemical bonds.
4. An electric current from the battery will also break down water to form hydrogen gas and hydroxide ions. In this activity the zinc ion is more readily reduced than water; thus, zinc metal is produced.
5. The larger the voltage of the battery, the faster the reaction will be.
6. A way to remember the cathode and anode reactions is the phrase, "a RED CAT and AN OX" (*red*uction occurs at the *cat*hode, and *ox*idation occurs at the *an*ode).

ANSWERS TO THE QUESTIONS

1. Evidence of reaction includes the formation of a gray coating on one electrode and the formation of crystals and an amber solution around the other electrode.
2. They are similar to the original materials (zinc and iodine) in color.
3. The battery provided the electrical energy required to decompose the compound.

25. Osmosis on a Small Scale

Osmosis is the movement of water through a semipermeable membrane from an area where the solute concentration is low to an area where the solute concentration is greater. Plant and animal cells control nutrient concentrations by osmosis; pickles are made from cucumbers by osmosis; and venous blood plasma enters capillaries because of the effect of osmosis. In this activity you will carry out a reaction that results in the formation of a semipermeable membrane, and you will see how osmosis works.

Materials

1. Small beaker.
2. Copper sulfate ($CuSO_4$) solution, 0.1 M.
3. Crystals of sodium hexacyanoferrate(II) [$Na_4Fe(CN)_6$].

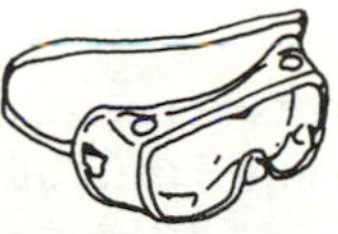

Procedure

1. Place about 50 mL of copper sulfate solution in a small beaker.
2. Drop in a few crystals of sodium hexacyanoferrate(II).
3. Record your observations.

Reaction

Copper sulfate reacts with sodium hexacyanoferrate(II) to form copper hexacyanoferrate(II), which forms a brown, semipermeable membrane:

$$2Cu^{2+}(aq) + [Fe(CN)_6]^{4-}(aq) \longrightarrow Cu_2Fe(CN)_6(s)$$

The copper sulfate solution is very dilute, only 0.1 M (about 2.5%). Therefore, the concentration of solute [sodium hexacyanoferrate(II)] inside the semipermeable membrane is much greater than that of solute (copper sulfate) outside the semipermeable membrane. Thus, water from the copper sulfate solution enters the membrane.

When water enters the membrane, the membrane swells and eventually bursts, throwing out the concentrated sodium hexacyanoferrate(II) solution. This solution immediately reacts with the copper in solution to produce more copper hexacyanoferrate(II), resulting in solid brown streamers and membranes.

Questions

1. What is a semipermeable membrane?
2. What is osmosis?
3. How is osmosis responsible for water entering a cell? (Hint: The concentration of solute, especially protein, is much greater inside the cell than outside.)

Notes for the Teacher

BACKGROUND

Demonstrating osmosis on a large scale can be done in several ways, including putting a raw egg (in its shell) in vinegar for 2 days or inserting a tube in a carrot. This activity uses a chemical reaction to produce a semipermeable membrane and then allows students to see, in a short period of time, what happens when solutions of different concentrations are on either side of the membrane.

MATERIALS

1. Copper sulfate solution ($CuSO_4$), 0.1 M. Dissolve 2.5 g of copper sulfate pentahydrate in 100 mL of water.
2. Sodium or potassium hexacyanoferrate(II) (also called sodium or potassium ferrocyanide).The hexacyanoferrate ion is very stable and does not release the toxic cyanide.

TEACHING TIPS

1. This activity works well as a demonstration. Put the solution in a Petri dish and project it with an overhead projector. Add a few crystals of sodium hexacyanoferrate(II) to produce the osmotic effect.
2. The pressure required to counteract the force of osmosis is called *osmotic pressure*.
3. Capillaries and plant roots act as semipermeable membranes and allow water and small molecules to enter cells and move through them.
4. Stress the fact that osmosis is a property of a solution in contact with a more concentrated solution.

ANSWERS TO THE QUESTIONS

1. A membrane that allows passage of water and small molecules.
2. The passage of water across a semipermeable membrane from a dilute to a more concentrated solution.
3. The concentration of solutes (e.g., proteins) in the cell is greater than the concentration of solute outside the cell. Thus, water enters the cell in an attempt to dilute the more concentrated solute. This process continues until an isotonic balance is reached between solutions inside and outside the cell.

26. Movement of Small Molecules and Water Through a Membrane

One of the basic processes in living systems is that of *osmosis*, the passage of water across a semipermeable membrane to an area of greater solute concentration. A related process is the simple *diffusion* of small molecules and ions across the membrane. Cells depend on these mechanisms to bring in nutrients and release metabolic products. In this activity we will look at evidence for both osmosis and diffusion.

Materials

1. Short piece of dialysis tubing (about 20–30 cm).
2. Starch solution.
3. Iodine solution.
4. Salt solution (1%).
5. Salt solution (10%).
6. Graduated cylinder.
7. Three beakers, 250 mL.

Procedure

1. Soften three pieces of dialysis tubing by holding them under running water and rubbing the ends between two fingers until the tubing separates.
2. Prepare three bags by tying a knot in one end of each piece of tubing.
3. Carefully measure 20 mL of 1% sodium chloride solution and add it to one bag.
4. Place this bag upright in a 250-mL beaker. Fill the beaker two-thirds full with 10% salt solution.
5. Measure 20 mL of 10% salt solution and place it in another dialysis bag.
6. Place this bag upright in another 250-mL beaker and add 1% salt solution to this beaker until the liquid level is the same as that in the first beaker. Set both beakers aside for 15 min.

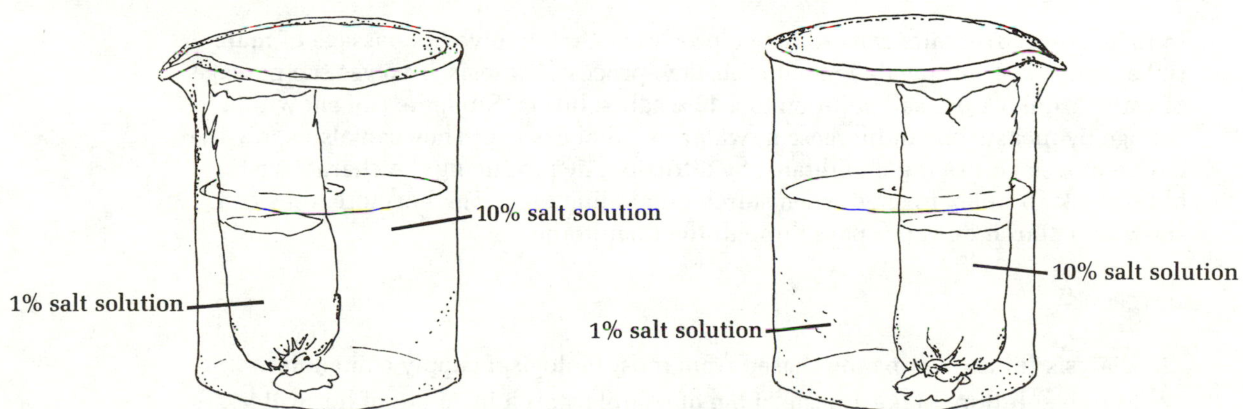

7. Add 25–30 mL of starch solution to the third dialysis bag. Rinse the outside of the bag.
8. Place the bag upright in another 250-mL beaker. Fill the beaker two-thirds full with water. Add 3 or 4 droppers of iodine solution. Observe the bag after 15 min or the next day.

9. Add a few drops of the iodine solution to a small amount of the starch solution in a separate test tube or beaker. Notice the blue-black complex. Set this tube aside.
10. After the allotted time, compare the levels in the first two beakers containing sodium chloride solutions. Has one decreased? Has the other increased? Why?
11. Remove the bags from the first two beakers. Carefully pour the solution from one bag into the graduated cylinder and record the volume. Repeat with the second bag. How does the volume of each bag compare with the original volume of the bag (20 mL)? Which decreased in volume? Which increased in volume? What happened to the starch solution inside the third bag?

Reactions

In the first two beakers, osmosis resulted in a movement of water from the less concentrated solution (1%) to the more concentrated solution (10%). This change increased the volume of water in the 10% solution bag and decreased the volume of water in the 1% bag.

Diffusion of iodine through the semipermeable dialysis tubing allowed iodine to react with starch to give the characteristic blue-black complex. The blue color is produced when the I_5^- or I_3^- complex fits inside the coiled molecule of amylose. The large starch molecule will not diffuse through the membrane.

starch–iodine complex

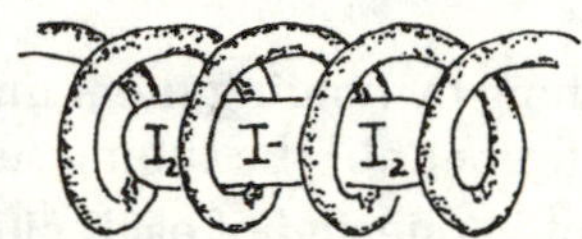

Questions

1. Which of these activities involved osmosis? Which involved diffusion?
2. How much water was gained or lost in each bag?
3. How do you know that iodine passed through the membrane?

Notes for the Teacher

BACKGROUND

In this activity students can study two processes that involve the passage of material across a semipermeable membrane. One process, osmosis, involves the passage of water from a 1.0% salt solution to a 10% salt solution. Students can show this change by measuring an increase in water in a dialysis bag. They can also show that iodine moves across the membrane, by diffusion, by producing the characteristic blue-black complex formed when starch and iodine react. Furthermore, they can show that starch does not pass through the membrane.

MATERIALS

1. Dialysis tubing can be purchased from most biological supply companies.
2. Starch solution. Make a paste of 5 g of soluble starch in 15 mL of hot water. Dissolve this in about 100 mL of boiling water. Cool the solution and dilute it to 1 L.
3. Potassium iodide–iodine solution. Dissolve 5.0 g potassium iodide, KI, in 500 mL water. Add 1.0 g iodine, I_2, and stir until dissolved. Place the solution in dropper bottles. Iodine is corrosive. Avoid touching the crystals.
4. Salt solutions: For 1%, dissolve 1 g sodium chloride, NaCl, in 99 mL of water. For 10%, dissolve 10 g sodium chloride, NaCl, in 90 mL of water.

TEACHING TIPS

1. The dialysis bag containing 10% salt solution placed in the beaker containing 1% salt solution will gain several milliliters of water. The bag containing the 1% salt solution placed in the beaker of 10% solution will lose several milliliters of water. The change in volume in the two beakers will be obvious.
2. The salt concentration in most human cells, which operate on the same principle as the membrane in this activity, is approximately 0.9%.
3. You might prefer to have students close both ends of the first two bags and weigh them before and after the experiment, rather than measure the volume. (Try both methods.)
4. The amount of water in the dialysis bag containing starch will also increase.
5. Both of these processes are important at the cellular level. *Osmotic pressure* (the pressure necessary to counteract the effect of osmosis) forces carbon dioxide and other products into venous blood and removes oxygen and nutrients from arterial blood. Osmotic pressure is also responsible for making pickles!
6. See Activity 25 for a very simple way to demonstrate osmosis.
7. You may prefer to use a 1% and 10% glucose solution (also called dextrose). Normal plasma glucose is about 5%.

ANSWERS TO THE QUESTIONS

1. The first two showed osmosis; the last (with starch) showed diffusion of iodine but also involved osmosis as water enters the bag also.
2. Answers will vary, depending on the length of time the bags were in the solutions.
3. The iodine reacted with the starch inside the bag to produce the characteristic blue-black complex.

Chemical Reactions

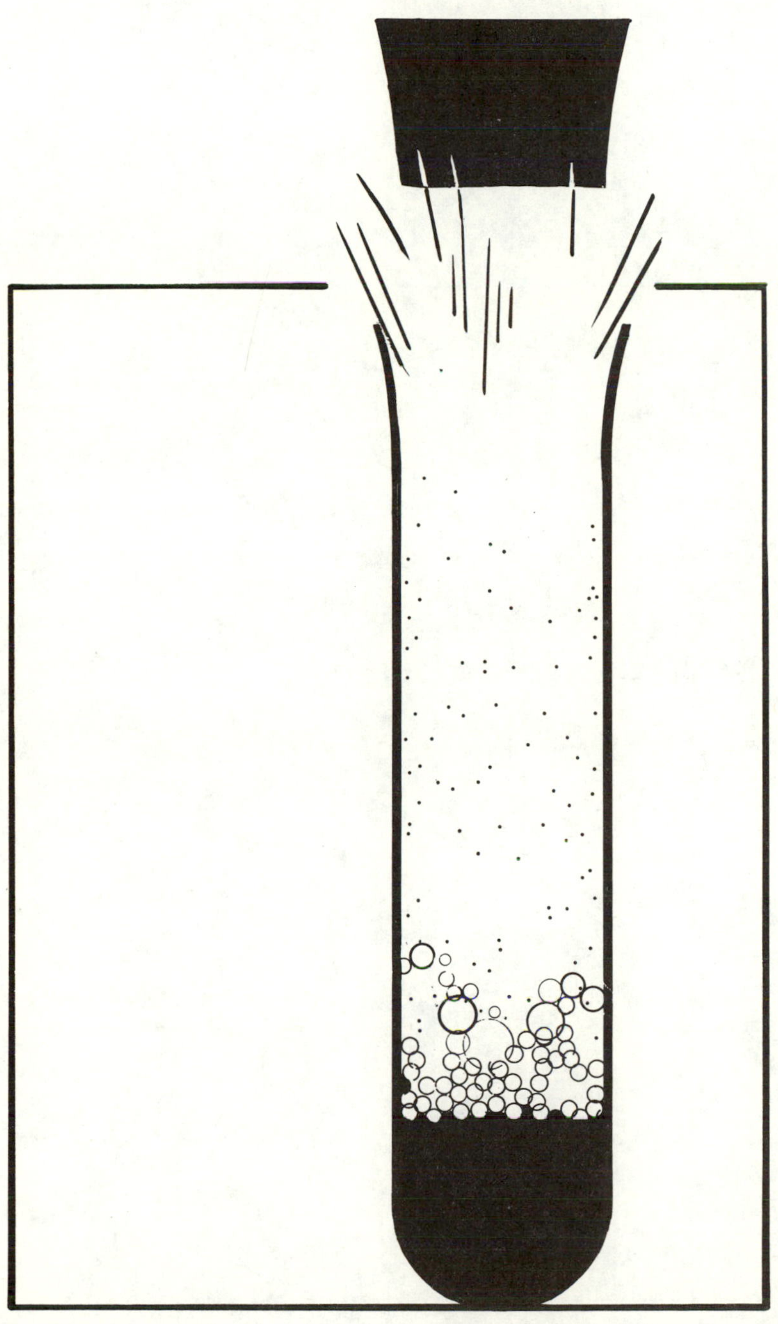

27. Floating Pennies

You will discover that pennies made since 1982 have a core of zinc that is plated with a thin layer of copper. Older pennies were made of copper that was uniform throughout. We will investigate the different properties of two metallic elements, zinc and copper.
This activity will take 2 days.

Materials

1. Hydrochloric acid (HCl), 6 M (about 20 mL).
2. One new penny (1983 or newer).
3. Beaker, 100 mL.

Procedure

1. Wear goggles.
2. Place 20 mL of 6 M hydrochloric acid in a 100-mL beaker. CAUTION! This reagent is a strong, corrosive acid. Wear gloves, goggles, and a face shield.
3. Scrape the edge of a new penny (1983 or newer) over concrete, coarse sandpaper, or a file to expose part of the zinc core.

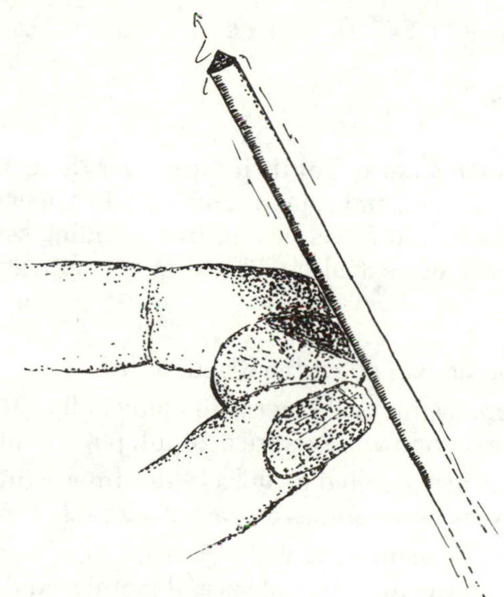

4. With forceps, carefully place the penny into the acid.
5. Leave the reaction undisturbed overnight.
6. When the penny is floating, carefully remove it with forceps. Rinse it in a container of water.

Reactions

1. The zinc reacts with the hydrochloric acid to produce hydrogen gas and dissolved zinc chloride:

$$Zn(s) + 2HCl(aq) \longrightarrow H_2(g) + Zn^{2+}(aq) + 2Cl^-(aq)$$

2. If the water is allowed to evaporate, the solid zinc chloride residue can be seen:

$$Zn^{2+}(aq) + 2Cl^-(aq) \longrightarrow ZnCl_2(s)$$

Questions

1. What is a physical difference between zinc and copper?
2. What is a chemical difference between zinc and copper in the presence of hydrochloric acid?
3. Describe a new substance that you see being produced in the reaction.
4. What observations do you make the next day? Suggest why the penny is floating.
5. Where do you think the zinc atoms have gone?

Notes for the Teacher

BACKGROUND

This activity provides an opportunity for students to develop several concepts: physical and chemical properties, reactions of metals and acids, production of hydrogen gas, use of safety precautions with acids, and conservation of atoms. Consider the extensions of this activity that are described later.

SOLUTIONS

Hydrochloric acid solution is 6 M. See Appendix 2 for instructions.

TEACHING TIPS

SAFETY NOTE: Caution students not to put their faces over the beaker. The reaction is exothermic (heat-producing), and some vapors of acid are evolved. Federal recommendations call for gloves and face shield or free-standing safety shield, as well as goggles, when using corrosive acids and bases in concentrations greater than 1 M.

1. The amount of zinc exposed will affect the reaction rate.
2. The amount of zinc in one penny will react with almost all of the acid in the amount specified. Each reaction will require a 20-mL portion of acid per penny.
3. Comparing the masses of new and old pennies is also interesting. Old pennies average 3.1 g each, whereas new pennies average 2.6 g each.
4. Vinegar, a dilute solution of acetic acid, will also react slowly with the zinc. If you cannot obtain HCl, scrape the edges of several pennies and place them in white vinegar. Because the concentration of acid in vinegar is only one-sixtieth of that of 6 M HCl, you will not observe all the zinc reacting, and the pennies will not become hollow.

ANSWERS TO THE QUESTIONS

1. Different colors.
2. Copper does not react. Zinc reacts to produce a gas (hydrogen). Alert students NOT to say "air bubbles are seen".
3. A gas is produced, as shown by the bubbles rising from the zinc.
4. The penny looks the same except that it floats. Hydrogen gas trapped in the penny causes it to float.
5. Zinc atoms are ions in solution, dissolved just like any salt. Try evaporating some of the solution on a piece of glass to see the zinc chloride salt residue.

28. "Magic" Color Changes in Solutions

When two solutions are mixed and a color change results, a chemical change has probably occurred. Such color changes often appear as if by magic and allow us to perform some nice chemical "tricks".

In this activity you will mix colorless solutions to form a pink solution, a clear solution, and finally another pink solution. The reactions are actually reactions between acids and bases, and indicators produce the color changes.

Materials

1. Four beakers or plastic glasses, 250 mL.
2. Phenolphthalein indicator solution.
3. Acetic acid (vinegar).
4. Household ammonia solution.

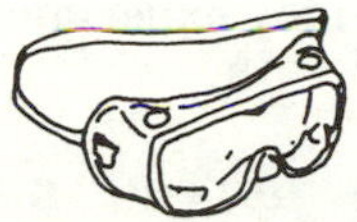

Procedure

1. Number the beakers 1, 2, 3, and 4.
2. Put 5 drops of phenolphthalein solution in beaker 1.
3. Put 5 drops of ammonia solution in beaker 2 and 15 drops in beaker 4.
4. Put about 10 drops of vinegar in beaker 3. Now you are ready to produce "magic" colors!
5. Add about 50 mL of water to each beaker. All beaker solutions should be colorless.
6. Note the appearance of the solution in beaker 1. Pour the contents from beaker 1 into beaker 2. Note the reaction.
7. Pour the contents of beaker 2 into beaker 3. Note the reaction.
8. Pour the contents of beaker 3 into beaker 4. Note the reaction.

Reactions

These reactions are all acid–base reactions in which an indicator is used.

1. Beaker 1 poured into beaker 2: Basic ammonia (NH_3) in beaker 2 reacts with the phenolphthalein in beaker 1 to turn it from colorless to pink.
2. Beaker 2 poured into beaker 3: The acid (vinegar) neutralizes the ammonia, and then extra acid turns the indicator colorless again. When ammonia is neutralized by vinegar, the new substances that are made are water and a salt, ammonium acetate.
3. Beaker 3 poured into beaker 4: The ammonia solution again neutralizes the acid, and extra ammonia produces a basic solution, which turns the indicator pink again.

Questions

1. What is an indicator?
2. Can you make the solution turn colorless again?
3. What kind of solution makes phenolphthalein turn pink?
4. Write word or chemical equations for each of the neutralization reactions.

Notes for the Teacher

BACKGROUND

This activity is a simple, yet fun, way to introduce acids, bases, and indicators. *Indicators* are compounds that react differently in acids and bases. You can make simple indicators from red cabbage juice, some fruits, and colored-petal flowers. The indicator phenolphthalein is colorless in acids but turns pink in bases.

SOLUTIONS

1. Phenolphthalein is a common acid–base indicator. It is also the active ingredient in Ex-Lax and can be obtained from a crushed tablet by soaking the tablet in any alcohol. Phenolphthalein is considered toxic.
2. Any dilute acids and bases will work. Students enjoy experimenting with household substances.

TEACHING TIPS

1. You might like to use household products, for example, vinegar and ammonia, for this activity.
2. Vary the amounts until you get the best color changes.
3. You may want to do this activity as a demonstration. However, this activity is fun for students to do directly and is safe.

ANSWERS TO THE QUESTIONS

1. An *indicator* is a substance that changes color in an acid or a base.
2. Yes, by adding an acidic solution.
3. A base turns phenolphthalein pink.
4. Reaction 1: Change of indicator by the presence of a base. When the indicator is colorless, its formula is HIn. When the indicator is pink, its formula is In^-. HIn and In^- refer to a large molecule called phenolphthalein.

$$\underset{\text{colorless}}{HIn(aq)} + \underset{\text{base in excess}}{OH^-(aq)} \longrightarrow H_2O(\ell) + \underset{\text{pink}}{In^-(aq)}$$

Reaction 2:

$$\underset{\text{base}}{OH^-(aq)} + \underset{\text{acid in excess}}{H^+(aq)} \longrightarrow H_2O(\ell)$$

Reaction 3: Excess base reacts with phenolphthalein as in reaction 1.

$$\underset{\text{acid}}{H^+(aq)} + \underset{\text{base in excess}}{OH^-(aq)} \longrightarrow H_2O(\ell)$$

29. Patriotic Solutions

There is no "magic" in chemistry. Everything that we observe in the laboratory, mysterious as it may seem, has a chemical explanation. In this activity we will mix three solutions and obtain surprising results. Then, we will examine the chemical reactions that produced the results.

Materials

1. Potassium thiocyanate (KSCN) solution, 10%.
2. Silver nitrate ($AgNO_3$) solution, 10%.
3. Potassium hexacyanoferrate(II) [$K_4Fe(CN)_6$)] solution, 10%.
4. Ferric chloride ($FeCl_3$) solution, 20%.
5. Three small beakers.
6. One large beaker or flask.

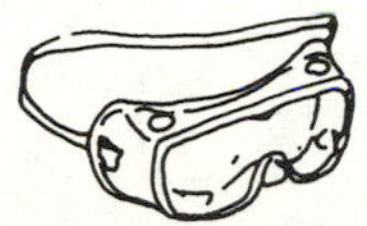

Procedure

1. Label three small beakers 1, 2, and 3.
2. Place 10 drops of potassium thiocyanate solution in beaker 1.
3. Place 10 drops of silver nitrate solution in beaker 2. Be careful. Wear gloves. Avoid touching the solution. Silver nitrate is corrosive.
4. Place 10 drops of potassium hexacyanoferrate(II) solution in beaker 3.
5. Place 20 drops of ferric chloride solution in the large beaker or flask and fill it three-fourths full with water. Describe the diluted ferric chloride solution.
6. Carefully pour the solution from the large container into beaker 1, then beaker 2, and finally beaker 3. What do you observe?
7. To dispose of your used solutions, you may pour the contents of beaker 1 down the drain. Filter the contents of beakers 2 and 3. Pour the filtrate down the drain and flush with water. Place the filter paper and the precipitate in the solid waste container.

Reactions

These reactions involve the formation of new colored compounds when ions in the beakers react with the ferric chloride solution.

Beaker 1: The ferric ion (Fe^{3+}) reacts with the thiocyanate ion (SCN^-) in the beaker to produce the blood-red complex, $[Fe(SCN)]^{2+}$. This test is a chemical test for the presence of the Fe^{3+} ion.

$$Fe^{3+}(aq) + SCN^-(aq) \longrightarrow \underset{\text{blood red}}{[Fe(SCN)]^{2+}(aq)}$$

Beaker 2: The silver ion (Ag^+) in the beaker reacts with the chloride ion (Cl^-) in the flask to produce the white solid, silver chloride (AgCl). This test is the chemical test for the presence of silver.

$$Ag^+(aq) + Cl^-(aq) \longrightarrow \underset{\text{white}}{AgCl(s)}$$

Beaker 3: The hexacyanoferrate(II) ion, $[Fe(CN)_6]^{4-}$, in the beaker reacts with the ferric ion (Fe^{3+}) in the flask to produce a deep blue solid called *Prussian blue*. Its structure is very complex, but it is probably $KFe_2(CN)_6$. This test is also a characteristic test for the ferric ion.

$$K^+(aq) + [Fe(CN)_6]^{4-}(aq) + Fe^{3+}(aq) \longrightarrow KFe_2(CN)_6(s)$$

Questions

1. What colors were produced in this activity?
2. Describe the chemical reaction that produced each color.
3. If you had a compound that you suspected contained the ferric ion (Fe^{3+}), how could you prove it?
4. How could you show that a certain unknown compound contained the silver ion (Ag^+)?

Notes for the Teacher

BACKGROUND

This activity gives students a little practice with reactions, equations, and characteristic tests for certain metal ions in solution. When mixed in the order suggested, the solutions produce red, white, and blue solutions.

SOLUTIONS

Because only a few drops of each solution are needed per student, the solutions should be provided in small dropper bottles.

Prepare each solution by adding 10–20 g of the solid to approximately 100 mL of distilled water. (You must use distilled water for the silver nitrate solution because the tap water may contain enough chloride ions for a precipitate of AgCl to form.)

TEACHING TIPS

SAFETY NOTE: Silver nitrate is corrosive and will stain the skin. Ferric chloride is also corrosive. These solutions are concentrated and should be used with care. Wear gloves and goggles when preparing solutions.

1. You should encourage students to be neat and careful when using any solution.
2. The thiocyanate and hexacyanoferrate(II) used in these solutions are not the same as the very poisonous cyanide. The cyanate complex is stable unless treated with hot, concentrated acid.
3. The structures of these ions are somewhat complex. You might want to emphasize the conservation of atoms and charge in each reaction.
4. In Activities 34 and 36, students can use these distinctive tests for certain ions.
5. You might prefer to perform this activity as a demonstration and use larger containers.
6. Dispose of precipitates as directed in Procedure 7. Disposal of silver nitrate solutions requires the procedure in Appendix 4B. However, we suggest that you save the solution for future use.

ANSWERS TO THE QUESTIONS

1. The ferric chloride solution is light yellow. The colors in the beakers are red, white, and blue.
2. See the Reactions section.
3. You could add a small amount of suspected ferric ion to either a potassium hexacyanoferrate(II) solution to produce a blue solution or to a potassium thiocyanate solution to produce a red solution.
4. If you add a solution containing a chloride (e.g., ferric chloride or sodium chloride), then white silver chloride will precipitate.

30. Colorful Cafeteria Napkins: Making Some Dyes

Some solutions, when mixed, result in a product that has a characteristic color and that can be used to dye certain fabrics and other materials. In this activity you can produce some of these dyes and use them to add color to white paper napkins.

Materials

1. Several white cafeteria-type napkins.
2. Ammonium iron(III) sulfate [$NH_4Fe(SO_4)_2 \cdot 3H_2O$] solution.
3. Potassium hexacyanoferrate(II) [$K_4Fe(CN)_6 \cdot 3H_2O$] solution.
4. Tannic acid ($C_{76}H_{52}O_{46}$) solution.
5. Sodium carbonate (Na_2CO_3) solution.
6. Cobalt chloride ($CoCl_2$) solution.
7. Paper towels or newspaper for blotting.

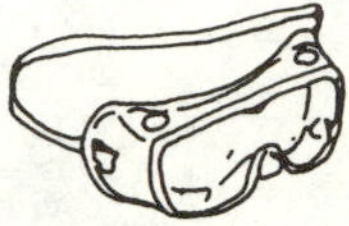

Procedure

1. Place about 50 mL of ammonium iron(III) sulfate solution in a 250-mL beaker.
2. Fold three white paper napkins several times and dip each napkin into the solution. Leave each napkin in the solution until about one-third of the napkin has soaked up the solution.
3. Remove each napkin and blot each one with paper towels to remove the excess solution. At the top of one napkin, write BLUE; at the top of another write BROWN; and at the top of another write BLACK. Place the napkins somewhere to dry.
4. Place about 50 mL of potassium hexacyanoferrate(II) solution into another 250-mL beaker.
5. Fold another napkin and dip it into the solution in the second beaker. When about one-third of the napkin has absorbed solution, remove the napkin, blot it with paper towels, and write GREEN at the top. Place this napkin in a place where it can dry also.
6. Prepare four 250-mL beakers by labeling them BLUE, BROWN, BLACK, and GREEN. Add about 50 mL of the following solutions to the corresponding beakers:

BLUE: potassium hexacyanoferrate(II) solution.
BROWN: sodium carbonate solution.
BLACK: tannic acid solution. This solution is toxic. Wear plastic gloves.
GREEN: cobalt chloride solution. This solution is toxic. Wear plastic gloves.

7. When the napkins have dried, carefully dip each labeled napkin into the beaker labeled with the same color. Remove each napkin. Blot each napkin to remove excess solution and let it dry. Avoid getting solutions on your skin. Wash your hands.
8. Record your observations.

BE CAREFUL !!!

Reactions

1. Blue dye: The blue dye is called *Prussian blue* and is due to the precipitate $KFe_2(CN)_6$.

$$K^+(aq) + Fe^{3+}(aq) + [Fe(CN)_6]^{4-}(aq) \longrightarrow KFe_2(CN)_6(s)$$

This test is commonly used to test for the presence of the Fe(III) ion.

2. Brown dye: The brown dye is iron(III) hydroxide. It is formed from the hydroxide produced when the carbonate ion reacts with water.

$$CO_3^{2-}(aq) + H_2O(\ell) \longrightarrow HCO_3^-(aq) + OH^-(aq)$$

$$Fe^{3+}(aq) + 3OH^-(aq) \longrightarrow Fe(OH)_3(s)$$

3. Black dye: The black dye is iron tannate. This molecule is very large and complex.
4. Green dye: The green dye is cobalt hexacyanoferrate(II).

$$2Co^{2+}(aq) + [Fe(CN)_6]^{4-}(aq) \longrightarrow Co_2Fe(CN)_6(s)$$

Questions

1. Are these dyes permanent dyes? Can they be removed with water? (Try it!)
2. Discuss the reactions that produce each color.
3. Try to devise reactions that will produce other colored dyes.

Notes for the Teacher

BACKGROUND

This activity gives students experience with chemical reactions that produce interesting colored products. In fact, these colored products have often been used as dyes because they are harmless and stable substances. Students can produce the dyes and use them to dye cotton materials or white paper napkins.

SOLUTIONS

The concentrations of the solutions are not critical in this activity. A 10% solution of each will work well.

Dissolve 100 g of each solid in 900 mL of distilled water. Stir until the solid completely dissolves.

TEACHING TIPS

1. Ammonium iron(III) sulfate is also called ferric ammonium sulfate.
2. Students should avoid contact with these (and any other) solutions. Provide plenty of newspaper and paper towels for blotting excess liquid.
3. Have students wash their hands thoroughly after mixing the chemicals and at the end of all laboratory work. Use plastic disposable gloves when you dispense tannic acid and cobalt chloride solutions.
4. Try varying the concentrations of the solutions.
5. You may not want to stress the reactions involved and the nature of the complex ions formed at this point.
6. Any of these solutions can be discarded by flushing them down the sink with copious amounts of water.
7. The complex $[Fe(CN)_6]^{4-}$ is stable unless treated with hot, concentrated acid.

ANSWERS TO THE QUESTIONS

1. Yes, they are "fast" and will not be removed by washing.
2. See the Reactions section.
3. Have students do some library research. Chemical handbooks and volumes on dyes and pigments in the art section might be useful.

31. "Slime": A Shimmery, Clear Fluid Polymer

A *polymer* is a very large molecule that is like a chain of smaller, often identical, molecules linked together. In this activity two clear solutions will be mixed in a paper cup. You will produce a fluid polymer. If green food coloring is added, the product will be like the commercial polymer, *slime*. You will examine the many properties of the polymer you synthesize.

Materials

1. Poly(vinyl alcohol), 4% solution.
2. Sodium borate, 4% solution.
3. Small paper cups and stir stick.
4. Food color (preferably green).
5. Plastic disposable gloves.

Procedure

SAFETY NOTE: Wear safety goggles and plastic disposable gloves.

1. Place 20 mL of poly(vinyl alcohol) solution in a paper cup.
2. Add 3 mL of sodium borate solution.
3. Using a circular motion, stir vigorously with a stir stick.
4. As the solution begins to solidify, continue to stir.
5. When a gel has formed, remove it and continue to shape it with your hands.
6. Investigate the different ways your polymer will act.
7. Write down your observations of the properties of the slime you have produced. Does it flow? Can it be flattened? Does it fracture (crack)? Does it get hot or cold when it flows? Will it bounce?

Reaction

When the sodium borate, $Na_2[B_4O_5(OH)_4]\cdot 8H_2O$, is added to the poly(vinyl alcohol), $[CH_2CHOH]_x$, a cross-linked polymer is formed. The new polymer is composed of strands of poly(vinyl alcohol) held together side-by-side (cross-linked) by the borate particles.

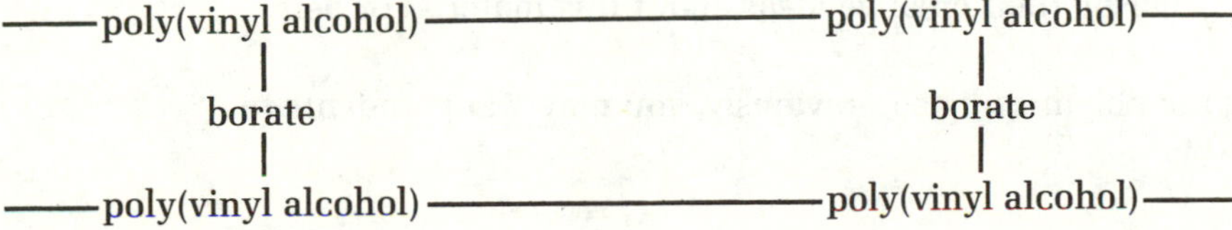

Questions

1. Write a descriptive paragraph about the polymer you have produced.
2. Name some polymers used to produce common substances.

Notes for the Teacher

BACKGROUND

This activity produces a simple cross-linked polymer that is representative of polymers in general. Between 30% and 50% of all chemists and chemical engineers work with polymers; thus, this class of chemicals holds an important place in industry. Synthetic polymers include silicone rubber, polyethylene, nylon, and poly(tetrafluoroethylene). Natural polymers include carbohydrates, proteins, and rubber.

SOLUTIONS

1. Poly(vinyl alcohol), approximately 4% solution: Add 40 g of solid poly(vinyl alcohol) to 1000 mL of water. Bring the water almost to a boil before adding the solid, small amounts at a time. If you have a magnetic stirrer, it would be useful. It may take 10–15 min for a clear solution to form. Avoid boiling the water. Avoid breathing the dust.
2. Sodium borate, 4% solution: Add 4 grams to 100 mL of water. This solution should also be warmed.
3. Green food coloring works well.
4. Wooden sticks or coffee stirring sticks work best.

TEACHING TIPS

SAFETY NOTE: This material should be handled carefully by students. Students should wash their hands after the experiment is completed and use plastic gloves when pouring the sodium borate solution.

1. Discard all paper cups and sticks so that no one will mistake them for drinking containers.
2. The students will discover various properties of this polymer. It can be flattened, and it will flow when held out at an arm's length.
3. Do not get slime on painted furniture or allow it to get into carpets. It will remove paint and stick to carpets.
4. The slime will last several days if it is kept in a closed jar. Mold will grow on it eventually.
5. The poly(vinyl alcohol) strands are linked by the borate particles through hydrogen bonding. A complete description of the theory may be found in *The Journal of Chemical Education* (Casassa, E. Z.; Sarquis, A. M.; VanDyke, C. H. *J. Chem. Ed.* **1986**, *63*(1), 57–60.

ANSWERS TO THE QUESTIONS

1. You may find students sharing their creative ideas about this material or being objectively descriptive.
2. In addition to naming the polymers listed previously, you may want students to do some library research.

32. Making a Super Ball

Two common liquids are mixed, and the reaction makes a solid that has the properties of rubber. The solid is a polymer. Chemists identify polymers as long chains in which molecules have linked up and twisted around each other, much like the paper chains people make for their Christmas trees. Nylon, polyethylene, wood (cellulose), proteins, and most plastics and rubbers are all polymers of particular molecules.

Materials

1. Sodium silicate (water glass) solution, 20 mL.
2. Ethyl alcohol, 5–10 mL.
3. Small paper cup for each solution.
4. Stirring stick.
5. Paper towels.

Procedure

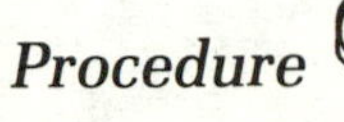

SAFETY NOTE: Wear safety goggles and plastic disposable gloves.

1. Measure 20 mL of sodium silicate solution and pour it into a small paper cup. Avoid contact with your skin.
2. Place 5 mL of ethyl alcohol in another paper cup. Alcohol is flammable. Extinguish all flames in the room.
3. Add the alcohol to the sodium silicate solution.
4. Using a circular motion, stir with the stick until the substance formed is solid.
5. Place the polymer in the palm of your hand and gently press with the palm of the other hand until a spherical ball that no longer crumbles is formed. Be patient. Discover a technique! Moisten the ball occasionally by holding it in a small stream of water from the faucet.
6. Bounce your ball!
7. Investigate as many property differences as you can between the two liquids and the solid polymer bouncy ball. Check the acid–base nature of each of the substances, as well as solubility in water and density.
8. Store the ball in a small plastic bag. If it crumbles, it can be re-formed.

Reaction

Silicon is a very interesting type of atom. Find its position on the periodic table of the elements. Like carbon, silicon makes four chemical bonds and can branch out in that many directions to make long chains. In sodium silicate, the silicon atom is bonded to four oxygen atoms and is not linked in any chains. The ethyl alcohol molecule is very simple and has just two carbon atoms. When sodium silicate and ethyl alcohol are put together, the silicate particles begin to link up with each other to form long chains as the ethyl groups (sometimes shown as "R") replace oxygen atoms in the silicate ion. Some become cross-linked between chains. Water molecules are byproducts of the formation of the polymerization bond.

The large molecule is a solid. It is a type of silicone polymer:

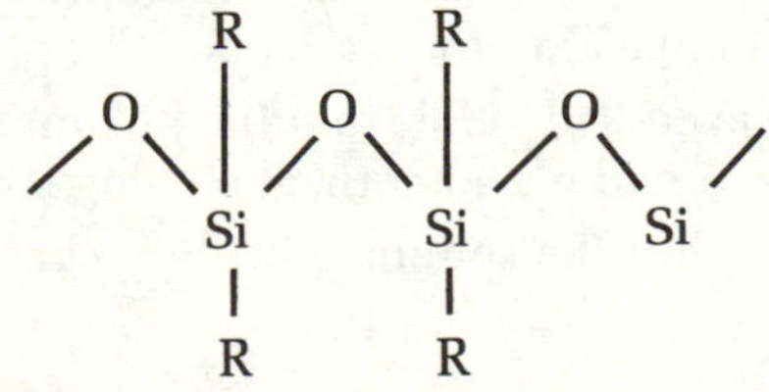

Questions

1. How did you know that a chemical reaction had taken place when the two liquids were mixed?
2. How could you find out what liquid was pressed out of the mass of crumbled solid as you formed the ball?
3. Compare your ball with those of the other members of the class. How many properties can you compare (e.g., diameter of sphere versus height of bounce)? List and compare them.

Notes for the Teacher

BACKGROUND

In this simple activity, students can easily and quickly make a rather complex polymer, silicone. Unlike carbon, silicon compounds are inorganic, and this polymer, unlike most other polymers, is an inorganic polymer. Sodium silicate consists of sodium ions paired with silicate ions as in $(Na^+)_2SiO_3^{2-}$. The silicate ion can also form a tetrahedral structure, SiO_4^{4-}, with silicon in the middle of the tetrahedron and an oxygen atom at each point of the tetrahedron. Long-chain polymers are formed by joining these tetrahedra at two corners of each to form negatively charged ions.

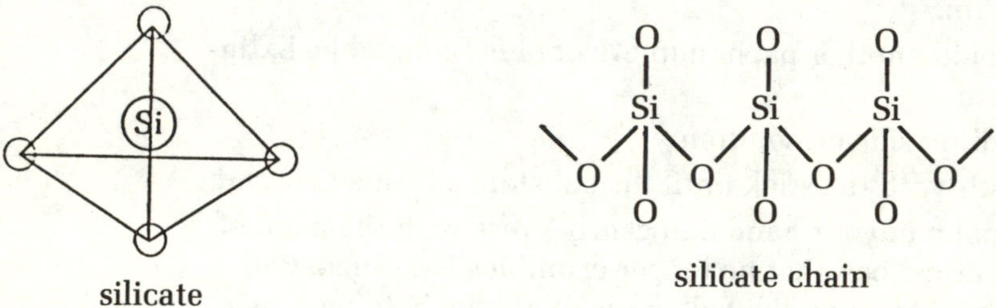

silicate silicate chain

When ethyl alcohol (CH_3CH_2OH) is added to sodium silicate, two oxygen atoms are replaced by the ethyl groups, and cross-linking of the silicate chains occurs with water also as a product. This process produces the rubberlike polymer.

SOLUTIONS

1. Sodium silicate solution: This solution is sold in many hardware stores as water glass.
2. Ethyl alcohol: 95% ethyl alcohol works best. You might try other alcohols including isopropyl alcohol.

TEACHING TIPS

SAFETY NOTE: Ethyl alcohol is flammable. Extinguish all flames in the room. Dispense flammable liquids in small containers, no more than 250 mL. Restrict the total flammable liquid in the classroom to 500 mL.

1. Playing putty (e.g., Silly Putty) is a mixture of silicone and chalk.
2. Some students may find that their polymers crumble when pressed. They should patiently press all of the material to form a ball.
3. For best results, gently press the excess alcohol from the polymer as it is rolled in the palms. Wet it occasionally to make a smooth, glistening surface.
4. Neither of these solutions is toxic, but the sodium silicate is corrosive. Wear gloves.

5. The type polymer produced in such reactions generally depends on the length of the -O-Si-O-Si- chains.
6. Silicones are resistant to temperature extremes and have been widely used as gaskets in spacecraft.
7. You can add a drop of food coloring to color the ball.

ANSWERS TO THE QUESTIONS

1. A solid immediately formed when the solutions were stirred.
2. The liquid is excess alcohol and perhaps a little water. You could collect the alcohol and burn it.
3. Balls will vary from smooth and round to rough and oddly shaped.

33. Flower Pigments as Acid–Base Indicators

Indicators are organic compounds that change colors at various pH values. They are useful to the chemist because most produce characteristic color changes at various degrees of acidity or basicity. Most indicators are organic dyes with complex formulas that are synthesized by reactions that are also complex. Some indicators, however, are easily derived from common plants. In this activity you will extract one of these natural indicators, use it to establish some color standards, and then use these standards to find the approximate pH values of substances found around the home. **This activity will take two class periods.**

Materials

1. A few colored flowers (e.g., roses, petunias, and geraniums). Dark-colored flowers will work best.
2. Ethanol (CH_3CH_2OH).
3. 10 test tubes.
4. Beaker.
5. Lemon juice.
6. White vinegar (acetic acid, CH_3COOH).
7. Boric acid (H_3BO_3) solution.
8. Sodium bicarbonate ($NaHCO_3$) solution.
9. Sodium carbonate (washing soda, Na_2CO_3) solution.
10. Borax (sodium borate, $Na_2[B_4O_5(OH)_4]\cdot 8H_2O$) solution.
11. Samples of materials for testing: household ammonia, limewater, soda water, wine, clear shampoo, antacids, and garden fertilizer.
12. Hot plate.

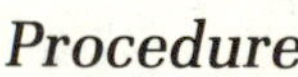

Procedure

1. Prepare the indicator.
   A. Place about 30 mL of ethanol in a small beaker.
   B. Add about 12 petals from the flower selected for the experiment.
   C. Place the beaker on a hot plate and gently heat for 5 min or until the color has faded from the petals. Stir the petals while they are being heated. CAUTION: Alcohols are flammable. Extinguish all flames in the room.
   D. Remove and discard the faded petals. The remaining liquid contains the acid–base indicator.

2. Prepare the color pH standards.
   A. Label eight test tubes as follows: tube a, pH 2; tube b, pH 3; tube c, pH 5; tube d, pH 7; tube e, pH 8; tube f, pH 9; tube g, pH 12; and tube h, pH 14.
   B. Add 10 mL of each of the following solutions to the corresponding test tube:
      a. Freshly squeezed and pitted lemon juice.
      b. White vinegar.
      c. Boric acid. Avoid contact with the skin.
      d. Water.
      e. Sodium bicarbonate (baking soda).
      f. Borax. Avoid contact with the skin.
      g. Sodium carbonate (washing soda).
      h. Sodium hydroxide (1.0 M). Avoid contact with the skin.
   C. Add 30 drops of the indicator solution to each test tube. Shake to mix thoroughly.

D. Notice the colors. Do they show a gradual change from a dark color (perhaps red) to a light color (perhaps yellow)?

3. Check out some unknowns.
   A. Prepare samples of materials collected around the house and laboratory. If solids are used, dissolve about 0.1 g per 10 mL of water.
   B. Place 30 drops of indicator solution in each test tube containing the substance with unknown pH. Shake the tube and compare the colors with the set of standards.
   C. Record your data and observations. Compare the colors produced by solutions with unknown pH values with the standards prepared in Procedure 2 to estimate their acidity or basicity. Compare your results with those of your classmates.

Reaction

Many flowers contain a pigment that belongs to a group of compounds called *anthocyanins.* This natural indicator is extracted with alcohol or water. It is red in very acidic solutions, violet in slightly acidic solutions, blue-green in slightly basic solutions, and yellow in very basic solutions.

Questions

1. Did your indicators show a gradual change from a dark color to a light color?
2. Why don't flower pigments dissolve when they become wet with rain?
3. Which flowers gave the best results?

Notes for the Teacher

BACKGROUND

The red and blue colors of most flowers and some vegetables are due to a group of organic substances known as *anthocyanins.* These compounds are soluble in alcohol and can be easily isolated and used as an effective acid–base indicator. This activity gives students experience with extracting a substance, preparing standards, and determining the approximate pH of unknown substances by comparing their results with the standards.

SOLUTIONS

For the most efficient operation, you should prepare the following solutions and simply label them a–h to correspond to the tubes that the students have labeled.

Tube	*pH*	*Solution*
a	2	Lemon juice. Should be freshly squeezed and pitted. Lemon juice has a high concentration of ascorbic acid.
b	3	White vinegar. Vinegar is acetic acid.
c	5	Boric acid. Dissolve 1 g of boric acid in 100 mL of water.
d	7	Water.
e	8	Sodium bicarbonate. Dissolve 1 g of sodium bicarbonate in 100 mL of water.
f	9	Borax. Dissolve 1 g of borax in 100 mL of water.
g	12	Sodium carbonate. Dissolve 1 g of sodium carbonate in 100 mL of water.
h	14	Sodium hydroxide solution is 1.0 M. Dissolve 4.0 g of NaOH in 100 mL of water. USE CARE: Do not touch the solid sodium hydroxide. It is caustic.

Other solutions that the students might test as unknowns include the following: wine (pH around 4), soda water (pH around 4), limewater (pH 11), household ammonia (pH 12), shampoos (most are near neutral pH, i.e., pH 7), garden fertilizers (most are slightly acidic), and various fruit juices (all will be acidic).

TEACHING TIPS

SAFETY NOTE: Borax and boric acid are toxic, and the solutions can be absorbed through the skin. Wear gloves. Sodium hydroxide (lye) is caustic. Wear gloves, goggles, and a face shield when you mix this solution.

1. Encourage students to share their results with their classmates. Perhaps a chart could be placed on the blackboard for data entry by the class.
2. Some flower pigments may not produce a gradual color change from dark to light (acidic to basic solution).
3. Flowers also contain other pigments. Of special interest are the carotenes and chlorophylls. Have students read about these photosynthetic pigments.
4. Red cabbage is an excellent source of anthocyanins. Simply pull off some leaves and heat them in water to extract the indicator solution. See Activity 47, "Acids and Bases in the Bathroom and Kitchen".
5. The color of the pigment may not be the same as the color of the flower.
6. Any of these solutions may be discarded by flushing them down the drain.
7. Try rubbing alcohol for the extraction.
8. Do not allow students to heat alcohol solutions directly with a burner. Use a hot plate.
9. The formula for borax (sodium borate) is $Na_2[B_4O_5(OH)_4]\cdot 8H_2O$.

ANSWERS TO THE QUESTIONS

1. Most indicators will show a gradual change from a dark color in acidic solution to a light color in basic solution. However, individual results may differ.
2. The pigment is most soluble in organic solvents such as alcohol or in hot water.
3. The darker flowers seem to give the most interesting indicators.

34. Precipitation: When Ions Find New Partners

When an ionic compound such as sodium chloride (NaCl) is dissolved in water, it separates into positive sodium ions (Na^+) and negative chloride ions (Cl^-). Thus, it is soluble in water. When silver nitrate ($AgNO_3$) is dissolved in water, it separates into silver ions (Ag^+) and nitrate ions (NO_3^-). However, if silver nitrate solution is added to sodium chloride solution, a white precipitate forms because the ions recombine to form new compounds. One compound, sodium nitrate ($NaNO_3$), is *soluble* in water and remains as sodium ions (Na^+) and nitrate ions (NO_3^-). The other combination of ions is silver chloride (AgCl). This compound is *insoluble* in water and thus forms a precipitate. It does not separate into silver ions (Ag^+) and chloride ions (Cl^-).

In this activity, you will mix some solutions of ions and note if a precipitate forms. If it does, you will identify the ions that make up the precipitate and give the precipitate a name. Jot down all possible recombinations of positive and negative ions on a piece of paper. You will also need to determine which compounds are soluble and which are not. Your teacher will help you with this determination.

Materials

1. Sodium chloride (NaCl) solution, 1%.
2. Silver nitrate ($AgNO_3$) solution, 1%.
3. Sodium iodide (NaI) solution, 1%.
4. Lead nitrate ($Pb[NO_3]_2$) solution, 1%.
5. Ammonium carbonate [$(NH_4)_2CO_3$] solution, 1%.
6. Calcium chloride ($CaCl_2$) solution, 1%.
7. Barium chloride ($BaCl_2$) solution, 1%.
8. 11 test tubes.
9. Droppers.

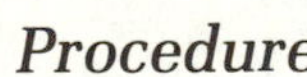

Procedure

SAFETY NOTE: Wear safety goggles and plastic disposable gloves.

1. You will need six test tubes to hold your stock solutions. Place about 5 mL of each solution listed in the Materials section in a test tube. Label each tube so that you will know what it contains. (Your teacher may want you to use small dropper bottles instead.) You will use these solutions in the following procedure. Avoid getting these solutions on your skin.
2. Add one full dropper of sodium chloride solution to a clean test tube. Write down the ions that are present in this solution.
3. Add, 1 drop at a time, a few drops of silver nitrate solution to the sodium chloride solution. Write down the ions that are present in the silver nitrate solution. What do you observe? Identify the precipitate.
4. Repeat with each of the following pairs of solutions. In each case, write down the ions that are present in each solution and deduce the name of any precipitate that forms. YOU MUST POUR ALL OF THE USED SOLUTIONS INTO CONTAINERS THAT YOUR TEACHER WILL PROVIDE. DO NOT POUR ANY INTO THE SINK. DO NOT GET THE SOLUTIONS ON YOUR SKIN. IF YOU DO, WASH QUICKLY.

   A. Sodium chloride (NaCl) and silver nitrate ($AgNO_3$). (How do you know that the precipitate is not sodium nitrate?)
   B. Sodium iodide (NaI) and lead nitrate [$Pb(NO_3)_2$].
   C. Ammonium carbonate [$(NH_4)_2CO_3$] and calcium chloride ($CaCl_2$).
   D. Sodium sulfate (Na_2SO_4) and barium chloride ($BaCl_2$).
   E. Sodium iodide (NaI) and barium chloride ($BaCl_2$).

Reactions

For each reaction, we will show only the ions that form a precipitate. The others do not really react, so we do not need to show them. The ions that do not react are often called *spectator ions.*

A. $AgNO_3$ and NaCl:

$$Ag^+(aq) + Cl^-(aq) \longrightarrow AgCl(s)$$

B. $Pb(NO_3)_2$ and NaI:

$$Pb^{2+}(aq) + 2I^-(aq) \longrightarrow PbI_2(s)$$

C. $CaCl_2$ and $(NH_4)_2CO_3$:

$$Ca^{2+}(aq) + CO_3^{2-}(aq) \longrightarrow CaCO_3(s)$$

D. $BaCl_2$ and Na_2SO_4:

$$Ba^{2+}(aq) + SO_4^{2-}(aq) \longrightarrow BaSO_4(s)$$

E. $BaCl_2$ and NaI: no precipitate forms

Questions

1. How do you know that the precipitate in "A" is not sodium nitrate?
2. Why doesn't a precipitate form in "E"?
3. How do you know that the precipitate in "D" is not NaCl?
4. What do you think would happen if silver nitrate solution were added to sodium iodide solution? TRY IT! Name the precipitate formed, if any.

Notes for the Teacher

BACKGROUND

This activity gives students an opportunity to mix some ionic solutions and predict the formation of precipitates. This activity also gives students practice in writing reactions between ions in solution, using solubility rules, and developing general laboratory techniques.

MATERIALS

Prepare very dilute solutions of the following (try about 5 g of each solid per 500 mL of water):

1. Sodium chloride (NaCl). (Use ordinary table salt.)
2. Silver nitrate ($AgNO_3$). (Use only distilled water. Do not get the solid or solution on your skin.) CAUTION: Corrosive.
3. Sodium iodide (NaI).
4. Lead nitrate [$Pb(NO_3)_2$]. CAUTION: Toxic.
5. Ammonium carbonate [$(NH_4)_2CO_3$].
6. Calcium chloride ($CaCl_2$).

7. Sodium sulfate (Na_2SO_4).
8. Barium chloride ($BaCl_2$). CAUTION: Toxic.

TEACHING TIPS

SAFETY NOTE: Plastic gloves will protect students from contacting the solutions of silver nitrate, lead nitrate, and barium chloride.

1. You might prefer to provide these solutions in sets of dropper bottles. Each set could be used by 6–8 students, and it could be used several times. Spillage is then less likely. Soluble salts of lead, silver, and barium are considered toxic.
2. Provide waste cans for disposal of the precipitates from this activity. See Appendix 4B for disposal procedures.
3. Have sample solutions of the following available to show students that the ionic compounds are soluble: sodium chloride solution, sodium nitrate solution, and ammonium chloride solution.
4. Be especially careful when using silver nitrate. If this compound is spilled on the skin, instruct students to wash it off immediately. It will stain the skin, leaving brown marks. If a spill occurs, add a dilute solution of potassium iodide (KI) and then a dilute solution of potassium or sodium thiosulfate ($K_2S_2O_3$). This combination will remove the stain.
5. Calcium carbonate is a form of chalk.
6. Because barium sulfate is not soluble, it is used as a "barium milkshake". Patients are given this compound, and it shows up as a dark area on X-rays. It outlines the intestinal system and helps physicians to locate abnormal areas.
7. Have students notice the different types of precipitates (e.g., slight, heavy, colored, and gelatinous).

ANSWERS TO THE QUESTIONS

1. Because sodium nitrate is soluble. (Show them the bottle labeled "sodium nitrate".)
2. Because both barium iodide and sodium chloride are soluble.
3. Sodium chloride is also soluble.
4. A precipitate (silver iodide) would form. It would have to be this compound because we know that sodium nitrate is soluble.

35. Oxidation–Reduction in a Shake

Oxidation–reduction reactions are reactions in which electrons are transferred from one reactant to another. In this activity we will change copper ions to copper atoms by causing zinc atoms to give up electrons. Because copper ions are blue in solution and copper metal has completely different properties, we can easily follow this electron-transfer reaction.

Materials

1. Large test tube with stopper to fit.
2. Copper sulfate ($CuSO_4$) solution.
3. Zinc powder.

Procedure

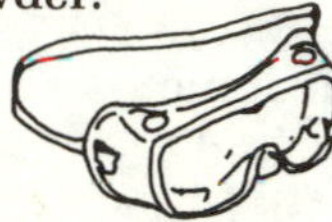

1. Fill the test tube two-thirds full with copper sulfate solution. Note the color of the solution.
2. Add 1 tsp of zinc powder to the tube. Note the appearance of the zinc metal as it settles to the bottom of the tube.
3. Stopper the tube firmly and shake it vigorously for 10–15 s.
4. Record your observations.

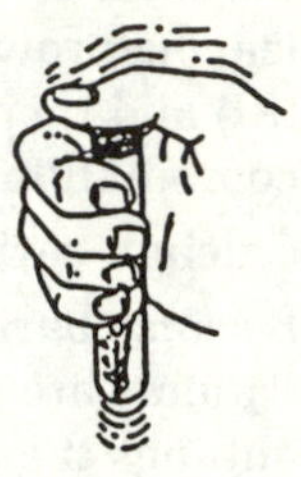

Reactions

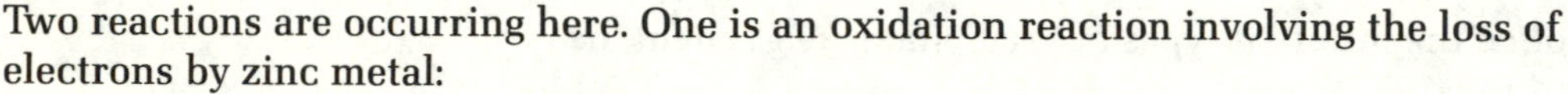

Two reactions are occurring here. One is an oxidation reaction involving the loss of electrons by zinc metal:

$$Zn(s) \longrightarrow Zn^{2+}(aq) + 2e^-$$

The other reaction is a reduction reaction or gain of electrons by copper ions:

$$Cu^{2+}(aq) + 2e^- \longrightarrow Cu(s)$$

We call this reaction *reduction* because the charge on the copper has been reduced from +2 to 0.

As a result of these two simultaneous reactions, the blue color of the copper ion diminishes, copper metal is formed, and zinc metal is changed to zinc ions, which go into solution.

The reaction also produces heat. Perhaps you noticed that the test tube became warm. Such reactions are called *exothermic* reactions.

These two reactions can be written as an overall reaction:

$$Zn(s) + Cu^{2+}(aq) \longrightarrow Zn^{2+}(aq) + Cu(s) + \text{heat}$$

Questions

1. What evidence indicates that a chemical reaction occurred?
2. Describe the oxidation reaction.
3. What happened to the copper ions? To the zinc metal?

Notes for the Teacher

BACKGROUND

This activity gives students experience with a simple oxidation–reduction reaction between metallic zinc and copper ions. Copper ions form a deep blue color in solution. When these ions react with metallic zinc, which is a more reactive metal than copper, the zinc atoms lose electrons to the copper ions to form copper metal. When all of the copper ions have left the solution, the blue will disappear.

SOLUTIONS

1. Copper sulfate solution: Make a 0.1 M solution by dissolving 25.0 g of copper sulfate pentahydrate ($CuSO_4 \cdot 5H_2O$) in 1 L of water.

   Zinc powder works well. You may be able to use mossy zinc if it is pounded into smaller pieces.

TEACHING TIPS

1. The large surface area of the zinc powder gives a rapid reaction. If you use zinc strips or large pieces, the reaction will take longer.
2. Copper sulfate is usually purchased as the pentahydrate: five water molecules are attached to each copper sulfate.
3. The tube should get warm, the blue color should diminish, and the gray metallic zinc should be replaced by shiny metallic copper. See Activity 42 for a quantitative study of this reaction.
4. The reactant that contains the atom that is reduced (Cu^{2+}) is called the *oxidizing agent*. The reactant that contains the atom that is oxidized (Zn) is called the *reducing agent*.
5. The oxidizing agent is the electron acceptor; the reducing agent is the electron donor.
6. This reaction is actually a very old one. Alchemists in the 1600s performed this experiment and thought that they had magically changed one metal into another.

ANSWERS TO THE QUESTIONS

1. Heat was evolved, a color change occurred (blue to colorless), and the product (copper metal) was unlike that of the reactant (zinc metal).
2. The oxidation reaction involved the loss of two electrons by zinc metal, which changed it into a zinc ion (Zn^{2+}).
3. Copper ions accepted two electrons to form copper metal. Zinc metal lost two electrons to form zinc ions. The zinc ions are soluble, and they form a colorless solution.

Chemical Energy and Rates of Reaction

36. T-Shirt Designs by "Chemotype"

Create your own T-shirt design and then repeat a time-honored procedure to produce a "positive" design from a "negative" design by using a chemical process similar to the process for making blueprints.

Materials

1. Black and white design transferred to an overhead transparency by photocopying, photographic negative, or pattern cut from heavy paper or aluminum foil. Two glass plates of an appropriate size.
2. Iron(III) ammonium citrate.
3. Potassium hexacyanoferrate(III) [$K_3Fe(CN)_6$].
4. White untreated 100% cotton fabric or T-shirt.

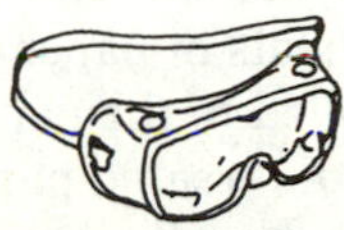

Procedure

1. Prepare your design by copying a picture onto an overhead transparency by photocopying or create a design by hand with an overhead transparency pen. You may choose to use a photographic negative, which will also work, although most negatives are small.
2. Prepare the fabric by smoothing it tautly and protecting the rest of the T-shirt with plastic.
3. Prepare the solutions by dissolving 30 g of iron(III) ammonium citrate in 100 mL of water (solution A) and 30 g of potassium hexacyanoferrate(III) in 100 mL of water (solution B).
4. When ready to print, mix equal volumes of solution A with solution B. The room light should be as dim as possible.
5. Paint the mixture onto the design area of the fabric with a brush, sponge, or cotton balls (wear protective gloves).
6. Allow to dry in a dark place.
7. When the fabric is dry, place a piece of picture glass under the design area.
8. Place the design on the treated area.
9. Cover with another piece of glass.
10. Expose the print to direct strong sunlight for 10–30 min.
11. Rinse in lukewarm water.
12. Allow your design to dry.
13. To wash the fabric, use mild soap.

Reactions

If you used an overhead transparency, you created a negative design from a positive design because the light reacted with the exposed chemicals it fell upon. In the reaction a photosensitive compound, iron(III) ammonium citrate, absorbs solar energy to cause an electron to be transferred to one of the iron(III) ions:

$$Fe^{3+} + e^- \longrightarrow Fe^{2+}$$

The ferrous ion (Fe^{2+}) reacts with the hexacyanoferrate(III) ion to form an insoluble blue compound:

$$K^+(aq) + Fe^{2+}(aq) + Fe(CN)_6{}^{3-}(aq) \longrightarrow KFe_2(CN)_6(s)$$

Questions

1. Why must the procedure be kept away from any light until the entire package is put together?
2. Why is this reaction called the blueprint reaction?
3. What other photochemical reactions do you know about?

Notes for the Teacher

BACKGROUND

A *blueprint* is an exact copy of the plans for a building or piece of machinery. Before modern copiers were invented, a chemical process for making blueprints was practiced for many generations and was the most efficient and useful way to produce many copies of the large sheets of paper on which plans were usually drawn. Solutions A and B were applied to the paper, and the original drawing on translucent paper was placed over the treated paper. After a bright light was shone on the papers, the reaction that was described took place and provided white lines on a blue background. Today, with modern copiers, blueprints are created by a process that produces black or blue lines on white paper.

TEACHING TIPS

1. This reaction is a fascinating example of the oxidation–reduction of iron.
2. This reaction also illustrates the idea of energy absorption by electrons (in citrate) to cause bond breakage and electron transfer. Iron(III) ions are reduced by the citrate to form iron(II) ions. The iron(II) ions, in turn, react with the hexacyanoferrate(III) ions to form the blue precipitate.
3. The concentrations of the solutions are not important, although the more concentrated the solutions are, the more blue product that will form.
4. A 6- × 10-in. design area will require between 20 and 30 mL of each solution.
5. Work in a room with diffused light from the windows and no artificial lights, if possible. Working in the dark is not necessary, although the treated fabric should be allowed to dry in the dark (e.g., inside a cardboard box).
6. If the sun is bright, the reaction may occur in 5–10 min. The reaction will also take place, although more slowly, on a cloudy day. Students might experiment with ultraviolet lamps or even fluorescent lights.
7. Be sure to rinse out the solutions well.
8. Large professional-quality negatives, such as those used by portrait photographers, make nice designs for quilt blocks or for a fine cotton print to be framed.
9. The exact formula of ammonium iron(III) citrate has not been determined. Actually, there are two compounds: one, with 14.5–16% iron, is green; the other, with 16.5–18.5% iron, is brown!

ANSWERS TO THE QUESTIONS

1. The reactions will take place if light falls on the combined chemicals.
2. The background is blue.
3. Photosynthesis, photography (i.e., silver halides on black-and-white film and dyes on color film), and reactions of chemicals in the eye during the vision process.

37. Lemon Voltage

You can increase the power of your lemon! Try two different metal strips and a lemon to cause a flow of electrons. We will know if a flow of electrons occurs because electrons will move the needle on a voltmeter, light a bulb, or ring a bell. We will discover the combination of metals that produces the greatest voltage. Voltage is a measure of how readily electrons flow from one electrode to the other.

Materials

1. Lemon.
2. Metal pieces such as a nickel, dime, penny, zinc strip, lead strip, magnesium strip, iron nail, and aluminum, all freshly cleaned with steel wool.
3. Wire leads with connectors.
4. Voltmeter or bulb and socket.

Procedure

1. Make two deep slits in the lemon several centimeters apart.
2. Select two different metals and insert them in the lemon.
3. Using wire leads, connect the metals to the bulb or to the poles of the voltmeter. If no reading occurs, reverse the poles.
4. Test as many combinations of metals as you can.
5. Which two metals seem to produce the greatest voltage when they are paired together?
6. Wash your hands thoroughly when you finish laboratory work.

Reactions

When two different metals are connected by wires and ionic solutions that will conduct electricity, electrons will flow from one metal to the other through the wire if the solutions are connected. With the lemon, in some cases, one metal (M) is oxidized, and a dissolved metal (M^+) is reduced:

$$M \longrightarrow M^+ + e^- \quad \text{oxidized}$$

$$M^+ + e^- \longrightarrow M \quad \text{reduced}$$

The dissolved metal is formed when the lemon acid reacts with the metal electrode. In other cases, one metal is oxidized, and at the other electrode hydrogen ions in the lemon are reduced:

$$M \longrightarrow M^+ + e^- \quad \text{oxidized}$$

$$2H^+ + 2e^- \longrightarrow H_2(g) \quad \text{reduced}$$

Oxidation means loss of electrons. *Reduction* means gain of electrons. The electrons flow through the wire and move the needle on the voltmeter or light the bulb. You can tell which way the electrons are flowing if you use pole-finding paper. See Activity 98.

Questions

1. Which two metals produce the greatest voltage when used together?
2. Use these two metals in combination with the others to rank the metals from "hardest to lose electrons" at the head of the list to "easiest to lose electrons" at the end. You need a starting point. Silver is the common metal that is most stable (hardest to lose electrons).

Notes for the Teacher

BACKGROUND

The lemon is a good wet cell. The pulp acts as a barrier and a salt bridge at the same time. The acid in the lemon reacts with some of the metal so that ions of the metal are in the solution surrounding the electrode. The acid also supplies hydrogen ions, which are more spontaneously reduced than some of the metals. Alessandro Volta, an Italian scientist, devised the first battery in the 1790s by sandwiching salty wet cardboard between alternating pieces of zinc and copper. Perhaps he could have used a lemon or two just as easily.

TEACHING TIPS

SAFETY NOTE: Be sure no one eats the lemon. It will have dilute dissolved metals in the juice.

1. The more metals you have available, the more challenging this activity will be. You can probably borrow electrodes from the physics teacher.
2. If you use the same lemon for more than one class, make new slits for each class. You can also try apples, oranges, and potatoes. Discard the lemon or other fruit or vegetable used.
3. A table of reduction potentials in any chemistry book can be used to predict which combinations will give the greatest voltage. Surprisingly, the observed voltages in this activity are close to those predicted under standard conditions (where solutions are 1 M).
4. Oxide coatings on magnesium and aluminum will interfere with electrode reactions and thus lower the resulting voltage. The nickel is actually 90% copper and 10% nickel.
5. It might be fun to hook up several lemons in series to find the maximum voltage possible. Lemons hooked in this way form a battery. Which two electrodes should be used in each lemon? Can you connect enough lemon "batteries" to play a small 9-V radio? Try it!

ANSWERS TO THE QUESTIONS

1. The predicted voltage for silver reduction and magnesium oxidation is about 3 V. Without silver ions in solution, however, the reduced substance will be hydrogen, and the voltage will be about 2.4.
2. Silver > copper > hydrogen > lead > tin > nickel > iron > zinc > aluminum > magnesium.

38. Electrolysis I: Turning Water Molecules into Hydrogen and Oxygen Gases

Water molecules (HOH or H_2O) are made of hydrogen atoms bonded to oxygen atoms. Energy is needed to break these bonds so that the hydrogen atoms can form hydrogen molecules (HH or H_2) and the oxygen atoms can form oxygen molecules (OO or O_2). When electrical energy is used to break bonds, the process is called *electrolysis*. You will construct an electrolysis apparatus with a 9-V calculator battery or a 6-V battery, wire leads and connectors, and electrodes made of carbon pencil "leads". Also see Activity 39, "Electrolysis II", for more ideas.

Materials

1. Small beakers, plastic cups, baby food jars, or Petri dishes.
2. Two pencil leads: Purchase carbon graphite or carve away the wood from one side of a pencil an inch or two above the point.
3. Battery, 6 or 9 V.
4. Two insulated wires with clips for the poles of the battery and for the electrodes.
5. Sodium sulfate.
6. Acid-base indicator solutions such as litmus, bromothymol blue, phenolphthalein, or red cabbage juice.

Procedure

1. Clip pencil leads to two insulated wires. Then clip the free end of each wire to one of the poles of the battery.
2. Place water in a dish. Sprinkle in a little sodium sulfate to help carry the electrical current.
3. Place the carbon electrodes opposite to each other in the solution. Make observations about what is happening at each electrode. Can you explain what is happening?

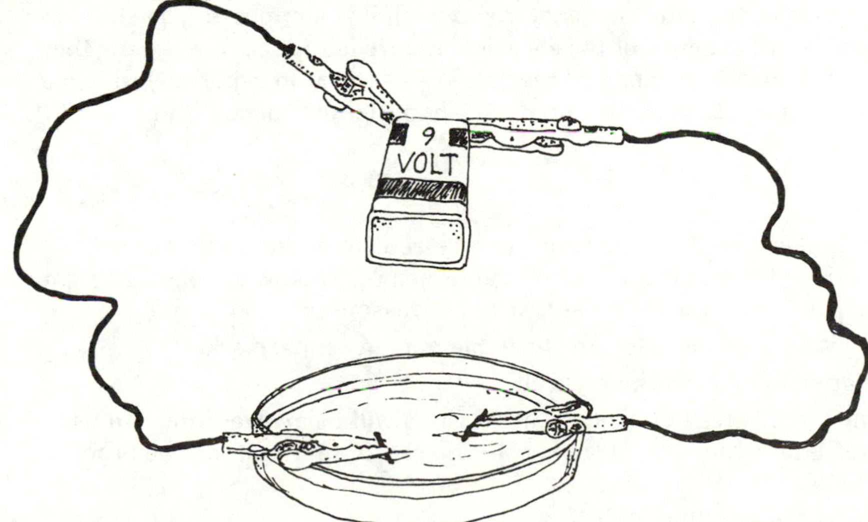

4. Stir the solution, then add several drops of the indicator to the dish of solution.
5. Place the electrodes in the solution as before. Make observations again. Can you explain what is happening?
6. Observe the color of the indicator solution under various known acidic and basic conditions. Compare with the color(s) seen when the water with indicator is electrolyzed.

Reactions

Oxygen gas and an acid are formed at one electrode. Hydrogen gas and a base are formed at the other. When the acid and the base meet in the dish, they form water! Meanwhile, the oxygen and hydrogen gases that form escape from the water. Thus, the electrolysis to break bonds in water is not so simple. It is many reactions.

Oxygen and acid formed at one electrode:

$$2H_2O(\ell) \longrightarrow 4H^+(aq) + O_2(g) + 4e^-$$

The H^+ is the acid. The e^- is the electric current that flows out of the solution and back to the battery. The O_2 is the oxygen.

Hydrogen and base formed at the other electrode:

$$4H_2O(\ell) + 4e^- \longrightarrow 2H_2(g) + 4OH^-(aq)$$

The electrons (e^-) are coming into the solution from the battery at this electrode. The H_2 is hydrogen gas. The OH^- is the base.

Questions

1. What gas is formed at the electrode where electric current flows into the solution?
2. What gas is formed at the electrode where electric current comes off the solution and flows back to the battery?
3. How could you collect the gases?
4. Predict which gas would be produced in twice the volume as the other gas.

Notes for the Teacher

BACKGROUND

Water is such a simple molecule that it serves as a great example of a compound with different properties than the elements from which it is composed. The electrolysis of water provides an example of the electrical nature of atoms, that is, how they accept and give up electrons in chemical reactions. This reaction is particularly nice because it is so easy and safe to perform with 9-V batteries and pencil lead.

TEACHING TIPS

1. This activity is also effective if it is done in a Petri dish by the teacher as a demonstration for the class on the overhead projector. This way the teacher can direct the discussion and help students with the reasoning.
2. The thicker graphite pencil leads are available in many supermarkets.
3. Calculator batteries (9 V) work very well.
4. Sodium sulfate works well because no other gases will be evolved. You can use table salt (NaCl), but you will get some chlorine gas. In that case, do the procedure in the hood.
5. Indicator colors are as follows:
   A. Litmus: blue with base and red with acid.
   B. Bromothymol blue: blue with base and yellow with acid.
   C. Phenolphthalein: pink with base and colorless with acid.
   D. Red cabbage juice: blue to green with base and red with acid.

6. Experiment with the effects of adding different salts to the water. See Activity 39.
7. Fill two small test tubes with water and hold them over the two electrodes to collect the escaping gases. This procedure requires several hands.

ANSWERS TO THE QUESTIONS

1. Hydrogen.
2. Oxygen.
3. See Teaching Tip 7.
4. Two volumes of hydrogen would be produced for each volume of oxygen.

39. Electrolysis II: Pushing Electricity Through a Salt Solution

When the two electrodes of a calculator battery are hooked to a circuit through a salt solution, interesting results can be seen. The battery does work on the salt particles, called *ions*, to add or remove electrons. Try to figure out what is being formed when the battery is hooked through several salt solutions.

Materials

1. Shallow container (plastic cup or Petri dish).
2. Samples of salts: sodium carbonate (Na_2CO_3), sodium chloride (NaCl), copper(II) chloride ($CuCl_2$), sodium iodide (NaI), and sodium bromide (NaBr).
3. Battery, 6 or 9 V.
4. Covered wires with alligator clips.
5. Two pencil leads (type for mechanical pencil used for art or type removed from wooden pencil).

Procedure

1. Add ¼ tsp of copper(II) chloride and a small amount of water in the dish to form a solution. Work in the hood with chloride and bromide solutions.
2. Attach one end of each wire to a battery terminal and the other end to the pencil lead. The pencil lead is graphite, a form of carbon.
3. Place the pencil lead electrodes in the salt solution and make observations. Keep the alligator clips dry. What do you see? By considering the name of the salt that you are using, can you figure out what you see?

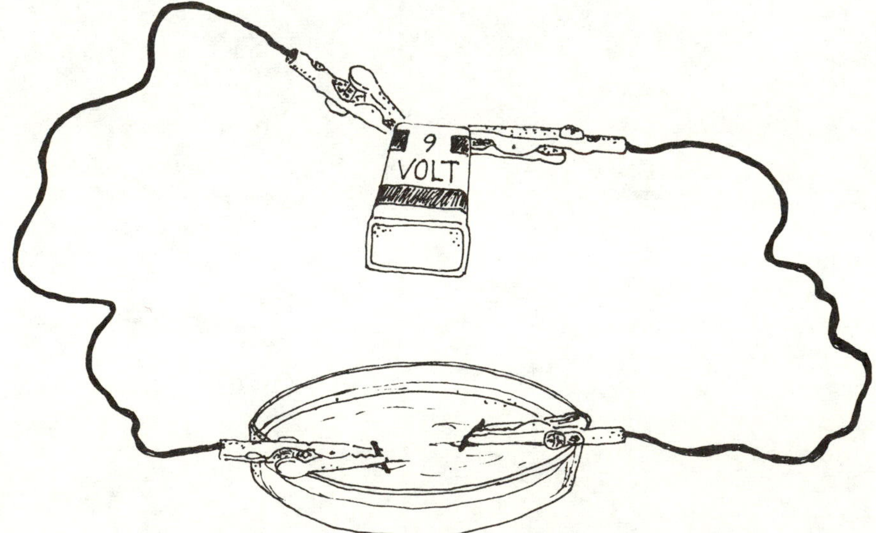

4. Clean out the container and prepare each of the other salt solutions. Repeat the electrolysis.

Reactions

Chemical energy is stored in the battery. When the battery electrodes are connected, electrons can flow and perform work such as running a calculator or lighting a light bulb. In these reactions, the electrons are added to something in the solution. For each electron added to something in the solution, an electron is removed from something else and flows back to the battery. For example, when a current is passed through a copper(II) chloride ($CuCl_2$) solution, the following reactions occur:

$Cu^{2+} + 2e^- \longrightarrow Cu$ (cathode reaction, electrons added)

$2Cl^- \longrightarrow Cl_2 + 2e^-$ (anode reaction, electrons removed)

Questions

1. What do you think forms at each electrode during electrolysis of the copper chloride salt solution?
2. What do you think forms during electrolysis of sodium iodide?
3. Can you identify any other products of the electrolysis of the other salt solutions?

Notes for the Teacher

BACKGROUND

See Activity 38, "Electrolysis I", for additional information about the process of sending electrons through solutions. The adding of electrons to dissolved metal salt solutions is the same process that is used industrially to "win" the metal from pulverized and dissolved metal ore. This process is the common way to obtain sodium metal, aluminum metal, magnesium metal, and a number of other metals from either dissolved solutions of their salts or from hot melted salts. This simple activity may initiate a discussion of the energy requirements of producing metals and the value of recycling.

TEACHING TIPS

SAFETY NOTE: Small volumes of chlorine or bromine gas will be produced. (See Answers to the Questions.) Use the hood.

1. See Electrolysis I for ideas about how to set up the apparatus and where to obtain the materials.
2. Experiment with various salts and try to guess the products of electrolysis. You will discover the relative stability of ions and atoms. For some clues about products, add a little phenolphthalein to the solution before you begin. In some cases the water is electrolyzed because the salt ions are more stable than the water; thus, the salt ions do not change. Phenolphthalein will identify the hydroxide ions (OH^-). If you see a gas forming, try to find a way to collect it. It may be hydrogen or oxygen from the water, or it may be another gas. See Activity 38. Show students how to detect an odor by fanning a small volume of air above the container toward the nose. Chlorine can be detected by odor.

ANSWERS TO THE QUESTIONS

1. Copper(II) chloride

   Negative electrode: copper metal (Cu).

   Positive electrode: chlorine gas (Cl_2).
2. Sodium iodide

   Negative electrode: hydrogen gas (H_2) and hydroxide ions (OH^-).

   Positive electrode: iodine dissolved in water. It is amber. (The iodide ion, when dissolved in water, appears as I_3^-).

3. • Sodium carbonate
   Negative electrode: hydrogen gas and hydroxide ions.
   Positive electrode: oxygen gas (O_2) and hydrogen ions.
   • Sodium chloride
   Negative electrode: hydrogen gas and hydroxide ions.
   Positive electrode: chlorine gas.
   • Sodium bromide
   Negative electrode: hydrogen gas and hydroxide ions.
   Positive electrode: bromine (Br_2). It is yellow.
   (The negative electrode is the *cathode*, and the positive electrode is the *anode*).

40. Dissolving Nails

Iron and iron compounds are found almost everywhere. In fact, 5.8% of the mass of the earth is iron. Since the Iron Age (around 1200 B.C.), when tools were first made of iron, people have tried to find ways to keep iron from reacting with oxygen in the air and with the acids and salts in soil and water. These reactions make rust and dissolved iron compounds and cause corrosion of the iron. You will study the appearance of dissolved iron in an acid solution in which nails sit for only 15 min.

Materials

1. 18 iron nails, "6 penny (6d)" or "8 penny (8d)".
2. Four beakers, 250 mL.
3. Hydrochloric acid, 0.1 M HCl, 400 mL in a 500-mL beaker.
4. Sodium chloride (NaCl), 17.4 g.
5. Sodium hydroxide (NaOH), 0.1 M.
6. Hydrogen peroxide (H_2O_2), 3%.
7. Potassium thiocyanate (KSCN), 0.2 M.
8. Iron color comparison standards.
9. Four test tubes.

Procedure

1. Pour 100 mL of 0.1 M HCl in one of the 250-mL beakers. Label this beaker "Control". Add six nails and allow them to react for 15 min.
2. Weigh about 17 g of sodium chloride and add it to the rest of the 0.1 M HCl solution.
3. Divide the NaCl–HCl solution into the three remaining beakers and add the nails: two in Beaker 1, four in Beaker 2, and six in Beaker 3.
4. Allow these beakers to react for 15 min.
5. At the end of 15 min, pour about 3 mL of acid solution from each of the beakers into small test tubes.
6. Add 10 drops of 0.1 M sodium hydroxide to each tube; this will partially neutralize the acid.
7. Add 3 drops of hydrogen peroxide to each tube to oxidize the iron.
8. Add 10 drops of 0.2 M potassium thiocyanate to each tube. What happens?
9. Compare the amount of color in each tube with the color comparison standards. How does the presence of salt affect the amount of dissolved iron? How does the amount of surface area affect the amount of dissolved iron?

Reactions

Corrosion (dissolving) of the iron becomes more rapid as we increase the surface area and as we increase the concentration of salts in solution. The salts assist in carrying the electric charge that occurs in this oxidation–reduction reaction.

$$Fe(s) + 2H^+(aq) \longrightarrow Fe^{2+}(aq) + H_2(g)$$

$H^+(aq)$ is the symbol for dissolved acid. $Fe^{2+}(aq)$ is the symbol for dissolved iron. Each iron atom has lost two electrons to the hydrogen. The iron is oxidized and the hydrogen is reduced. Hydrogen peroxide causes the iron to be oxidized further to Fe^{3+}.

When Fe^{3+} contacts the thiocyanate ion, SCN^-, a red complex ion is formed. The amount of red determines the relative amount of dissolved iron in the solution.

Questions

1. What do you observe as the nails react in the solution?
2. What do you observe as you add the hydrogen peroxide to the tubes?
3. How do you explain the difference between the control beaker and Beaker 3, both of which contain six nails?
4. How do you explain the differences among Beakers 1, 2, and 3, all of which contain the same solution?
5. In comparing your tubes with the color comparison standards, what do you find the concentrations of dissolved iron in your tubes to be?

Notes for the Teacher

BACKGROUND

The rate of any reaction can usually be increased if the concentration of the reactants is increased. Two rate factors are studied in this corrosion activity: an increased rate due to the increase in surface area of iron and an increased rate due to the increase in dissolved electrolyte. The sodium chloride provides increased pathways for the flow of electrons during oxidation and reduction. The students may be interested in corrosion of ships and metal parts in seawater. Preventing corrosion is one of the greatest challenges of the shipbuilding industry.

SOLUTIONS

1. Hydrochloric acid, 0.1 M; See Appendix 2.
2. Sodium hydroxide, 0.1 M: Dissolve 4.0 g in water to make 1 L of solution. Place the solution in small dropper bottles.
3. Hydrogen peroxide, 3%: Use the drugstore variety. It is stabilized.
4. Potassium thiocyanate, 0.2 M: Add 9.5 g to water to make 500 mL of solution. Place the solution in dropper bottles or in small beakers with droppers.
5. To prepare color standards, begin with a 1% iron chloride, $FeCl_3$, solution (1 g of iron chloride, $FeCl_3$, for 100 mL of solution). Remove 10 mL of this solution and dilute it with water to 100 mL. The second solution is 0.1%. Repeat the dilutions until the resulting solution is very pale. Place a sample of each dilution in a test tube of the size used by the students and arrange the tubes where all the students can examine them.

TEACHING TIPS

1. Each student should see the four experimental tubes compared; however, you may want to arrange the students in groups of four to save nails and beakers.
2. The students should observe the formation of hydrogen gas bubbles on the surface of the nails and the appearance of a green-yellow tint to the solution as the reaction proceeds.
3. Many manufacturers of breakfast cereal add reduced iron to their product so that the powdered iron will react with stomach acid, about 0.1 M HCl, to form dissolved iron, Fe^{2+}. You may place a white poly(tetrafluoroethylene) (Teflon) stir bar in a bowl of cereal on the magnetic stirrer for 10–15 min and pick up quite a bit of metallic iron.

4. Fe^{2+} is also easily oxidized by the air. Nails can be removed after 15 min of reacting, and the solutions can be exposed to the air until the next day. If the beakers are the same size, evaporation should be uniform.
5. See Activity 63, "Corrosion of Iron: The Statue of Liberty Reaction", for another example of how increased flow of electrons can increase the rate of iron reactions.
6. You can expect the iron concentrations in the solutions to range from 0.001 to 0.01%. Temperature and nail surface area and condition will cause differences in results.
7. You may want to study the concentrations of iron with a spectrophotometer. Consult the manual that is available with your spectrophotometer.

ANSWERS TO THE QUESTIONS

1. The nails are covered with tiny bubbles of hydrogen gas.
2. The solution in the tubes becomes yellow-green. This color is due to the dissolved iron reacting with the chloride from the acid.
3. The control beaker has no salt in it. The amount of dissolved iron is very small.
4. The more the surface area, the faster the reaction and the more product is formed.
5. Control: about 0.00001%; Beakers 1–3: about 0.0001–0.01%.

41. How Much Energy Is in a Nut?

We eat foods because they provide energy for our bodies. Determining how much energy different types of foods provide is important, especially if we want to plan a diet to balance the amount of energy we take in (eat) and the amount of energy we expend (work off). In this activity we will estimate the amount of energy contained in a pecan. To make this determination, we must release the energy in the nut by burning it and use the heat produced by the burning nut to heat a known quantity of water. By noting how much the temperature of the heated water increases, we can calculate how much energy the nut contains. For simplicity, we will assume that the nut is composed entirely of fat (actually, it is only about 85% fat).

Materials

1. Shelled pecan halves.
2. Paper clips.
3. Flask, 125 mL.
4. Small can (soup size) with top and bottom removed and air intake opening near the lower edge.
5. Thermometer.
6. Graduated cylinder, 100 mL.

Procedure

This activity will require that you make careful measurements of mass, volume, and temperature change. You must accurately record all of your measurements because you will use these data for calculations later.

1. Weigh the pecan half provided by your teacher. Record the mass.
2. Measure 100 mL of tap water and pour it in the 125-mL flask.
3. Record the temperature of the water in the flask.
4. Bend a paper clip, as shown in the sketch, to form a stand to support the pecan half when it is stuck into one end of the clip.
5. Arrange a setup as shown in the sketch.

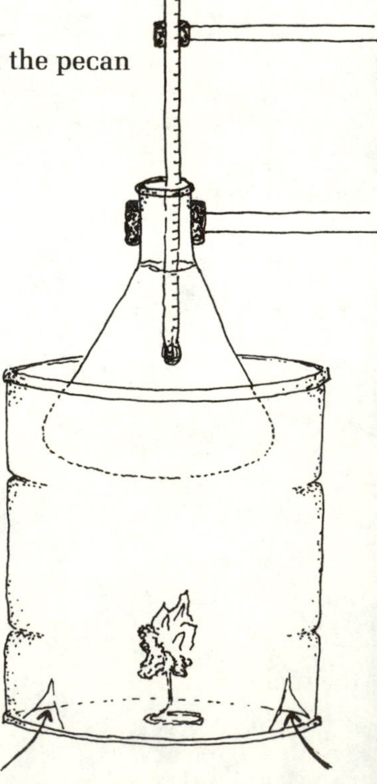

6. Light the pecan half with a match until it begins to burn. Immediately place the can over the pecan and its paper clip stand.
7. Loosen the clamp on the ring stand and place the flask inside the can so that it is close to, but not touching, the burning pecan. Tighten the clamp.
8. Place the thermometer in the flask and note the highest temperature the water reaches as the pecan burns.
9. Use the data you collected to calculate the amount of energy (heat) produced by the pecan.

Calculations

A convenient way to express heat is in calories. A *calorie* is the amount of heat required to increase the temperature of 1 g (1 mL) of water by 1 °C.

$$\text{calories} = \text{change in temperature (°C)} \times \text{volume of water (mL)}$$

For convenience, we also express the heat produced by a substance as the number of calories per gram. Thus, you can use this formula to convert the mass of your pecan piece and calculated number of calories to calories per gram of pecan:

$$\text{calories per gram of pecan} = \frac{\text{number of calories calculated}}{\text{mass of pecan piece}}$$

$$1000 \text{ cal} = 1 \text{ kcal}$$

A kcal (1000 cal) is the same energy as 1 Calorie (often called a "big" calorie) used by the home economist to determine food energy values.

Reaction

The composition of a pecan is mostly fats, about 85%. Pecans also contain a small amount of carbohydrate and some protein and about 3% water. The fat and fatlike material burn when the pecan is ignited. The products of this reaction are carbon dioxide, charcoal (carbon), and water.

Questions

1. How many calories of energy did your pecan piece produce in this activity?
2. What are some things that contribute to experimental errors in this activity?
3. According to your calculation, how many calories are produced per gram of pecan?

Notes for the Teacher

BACKGROUND

Scientists can easily determine how much heat energy a substance will produce when it burns. This technique is called *calorimetry*. In such an activity, the heat given off from a substance is used to heat water. If we know the mass of water and the increase in temperature, we can easily calculate the number of calories produced. This procedure permits us to compare various foods for their caloric value. Because our apparatus is simple, some of the heat is lost and our calculated values are not exact. Such deficiencies result in experimental errors.

CALCULATIONS

This activity allows students an opportunity to perform some simple calculations using the data they collected. We chose 100 mL of water because it makes calculations easier, but any volume can be used.

Sample data follow:

1. Mass of pecan piece: 1.2 g.
2. Volume of water: 100.0 mL. (Because the density of water is almost 1 g/cm^3, we can estimate that 1 mL will have a mass of approximately 1 g.)
3. Initial temperature of water: 25 °C.
4. Final temperature of water: 80 °C.

   Because the calorie is defined as the amount of heat required to increase the temperature of 1.0 g of water by 1 °C, we can calculate the total number of calories by simply multiplying the volume of water by the number of degrees that the temperature increased:

$$100 \text{ g} \times (1 \text{ cal/g °C}) \times 55 \text{ °C} = 5500 \text{ cal}$$

   This value is the number of calories produced by the sample, which weighed 1.2 g. If we divide the total number of calories by the mass of the pecan, we get the number of calories per gram of pecan:

$$\frac{5500 \text{ cal}}{1.2 \text{ g}} = 4583 \text{ cal/g} \approx 4600 \text{ cal/g (found experimentally)}$$

TEACHING TIPS

1. The can helps to direct heat from the burning nut onto the flask.
2. If students use a volume of water less than 100 mL, the water may boil, and they will not be able to record the highest temperature.
3. Ask students to put their data and calculations on the blackboard. Calculate averages for the class.
4. The burning nut produces a sooty flame. You may prefer to do this experiment in a hood or take students outside.
5. When oxidized (burned) in the body, 1.0 g of pecan will produce about 6800 cal (6.8 Cal). Have students look on Calorie tables to find the energy values of other foods.
6. Because the sooty flasks will need to be cleaned, you may prefer to use aluminum soft drink cans as the water containers. The soft drink can can be supported by the soup can with a stick or skewer through the soft drink can, and the thermometer can be placed in the hole in the top.
7. Try other foods. Brazil nuts and marshmallows work well.

ANSWERS TO THE QUESTIONS

1. Student answers will vary.
2. Such errors include loss of heat during the experiment, failure of the entire pecan to burn, the assumption that the nut is all fat (only about 85% of the pecan is fat), the absorption of some heat by the tin can and the glass flask, and poor measurement techniques.
3. Student answers will vary from 4 to 6 kcal per g.

42. The Heat Is On with Zinc and Copper Sulfate

When you add zinc granules to a solution of blue copper sulfate, a number of changes can be seen. If the amount of zinc is doubled, will the changes double? Predict the result of adding successively larger amounts of zinc to the copper sulfate solution. Your class will be a research team. Each group will perform part of the experiment and add their part of the results to the class results.

Materials

1. Granular zinc (Zn), assigned mass (1, 2, 4, or 8 g).
2. Copper sulfate solution, 50 mL.
3. Thermometer.
4. Polystyrene coffee cup.
5. Graduated cylinder.

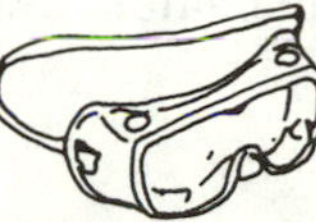

Procedure

1. Pour 50 mL of copper sulfate solution into a polystyrene coffee cup. Set the cup in a beaker to make it stable.
2. Place the thermometer in the solution.
3. Record the temperature of the solution.
4. Weigh out the number of grams of zinc assigned to you (1, 2, 4, or 8 g).
5. Stir the zinc into the solution with a plastic spoon.
6. Record the highest temperature reached during this reaction.
7. Without checking other students' data, predict the results of the addition of the other masses of zinc.
8. Collect all data from the class.
9. Construct a graph by plotting the amount of zinc on the *x* axis and the highest temperature reached on the *y* axis. Do the results match your prediction?

Reaction

In this reaction electrons are transferred from the zinc metal to the dissolved copper ions. The copper appears as brown copper metal atoms. The zinc atoms that have lost electrons then dissolve in the water. See Activity 35, "Oxidation–Reduction in a Shake".

$$Zn(s) + Cu^{2+}(aq) \longrightarrow Zn^{2+}(aq) + Cu(s)$$

The reaction is *exothermic*, that is, it gives off heat. The reaction takes place until either the zinc or copper sulfate is used up. The one that is used up first (the limiting reagent) determines the amount of heat given off.

Questions

1. Which substance is the limiting reagent in the mixture of 1 g of zinc plus 50 mL of copper sulfate solution?
2. Which substance is the limiting reagent in the mixture of 8 g of zinc plus 50 mL of copper sulfate solution? How do you know?
3. Explain the curve on your graph. Why is the temperature increase not regular as the amount of zinc is doubled?
4. What happened to the zinc atoms? What happened to the copper atoms?

Notes for the Teacher

BACKGROUND

Students often have difficulty seeing that one reactant runs out before the other(s) unless the chemist has very precisely weighed out each of the reactants. In this activity students can see the effects of varying one reactant so that it is limiting when weighed out in the smaller amount and in excess when weighed out in the larger amount. Students can track the course of the reaction by monitoring the temperature increase. By plotting the temperatures vs. mass of zinc used on a graph, students can see the limiting effect of each reagent in turn.

SOLUTIONS

1. Copper sulfate solution is 1.0 M: Add 250 g of copper sulfate pentahydrate ($CuSO_4 \cdot 5H_2O$) to water to make 1 L of solution.
2. Use granular zinc or zinc powder.

TEACHING TIPS

1. Assign the following amounts of zinc: 1, 2, 4, and 8 g.
2. If each trial is repeated by several groups, the results can be averaged.
3. When the students construct graphs, help them with the entries on the zinc axis so that the number increments are equivalent.

Sample graph:

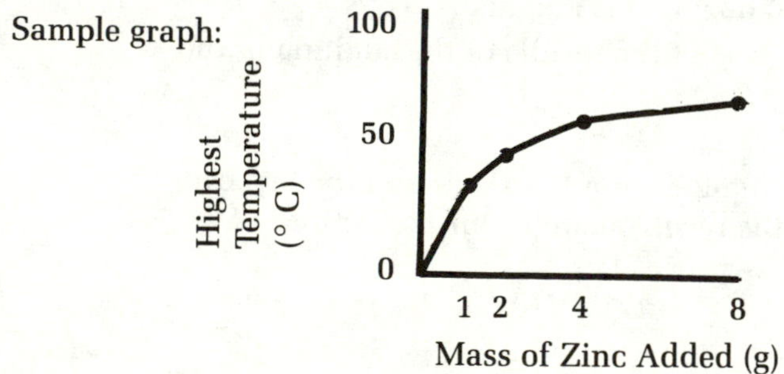

4. Destroy the polystyrene cups and spoons because they have been used with chemicals.

ANSWERS TO THE QUESTIONS

1. Zinc is the limiting reagent.
2. Copper sulfate is the limiting reagent. We can see that the copper sulfate is used up because the blue color is gone from the solution.
3. The reactions are faster and the temperature increases are greater as the amount of zinc increases up to about 3 g of zinc. The amount of copper sulfate stays the same. This condition then limits the extent of reaction and the temperature increase.
4. The zinc atoms lose electrons and dissolve in the solution. The copper atoms gain electrons and can be seen as metallic copper.

43. Evaporation and Temperature Change

How is temperature related to evaporation? When a substance evaporates, molecules must gain energy to leave the surface of the liquid. They get this energy from their surroundings. In this activity we compare how much heat is required for various substances to evaporate by measuring how much the surrounding temperature decreases.

Materials

1. Thermometer.
2. Several balls of cotton.
3. Small samples of acetone, ethyl alcohol, isopropyl alcohol, water, and perfume.

Procedure

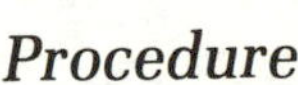

1. Twist the mercury-bulb end of the thermometer into a cotton ball to make a giant cotton swab.

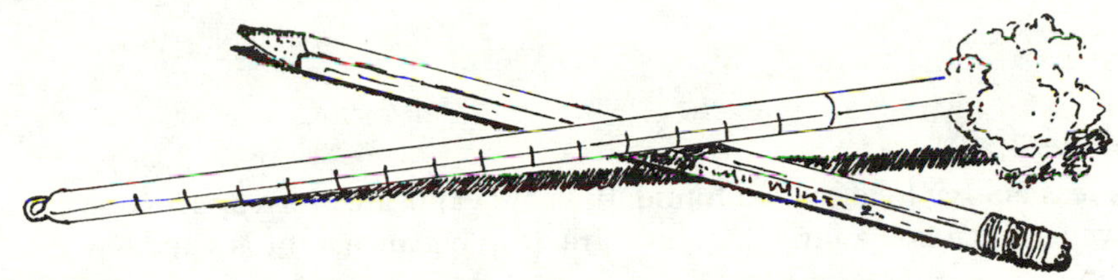

2. Place the thermometer with the cotton ball on the desk and prop up the cotton ball end with a pencil so that it does not touch the desk top.
3. Note and record the temperature of the thermometer.
4. Add enough alcohol to wet the cotton ball (about 10–15 drops).
5. Watch the thermometer to see if a change in temperature occurs. Note the lowest reading on the thermometer. Record your results.
6. Allow the thermometer to return to room temperature.
7. Using new cotton balls, repeat with acetone, isopropyl alcohol, water, and perfume.

Questions

1. Which substance caused the lowest temperature when it evaporated?
2. How is this result related to the motion of molecules?
3. List the materials in the order in which they evaporate.

Notes for the Teacher

BACKGROUND

When surface molecules of a liquid gain enough energy, they will overcome attractions of other molecules in the liquid and leave the surface. This process is *evaporation*. Those substances that require less energy evaporate more readily. We can smell perfume and acetone because they evaporate readily and their molecules activate odor-receiving cells in our noses. When evaporation occurs, the heat required for evaporation comes from the surroundings. With perfume, the heat needed for evaporation comes from the skin. This situation is why an "alcohol" rub feels so cool and comforting to the skin.

TEACHING TIPS

SAFETY NOTE: Acetone, alcohol, and perfume are flammable. Extinguish all flames in the room.

1. Provide samples of several liquids. Use small quantities to minimize hazards.
2. Stress data gathering and record keeping in this activity.
3. Be careful with thermometers. They are easily broken. If a mercury thermometer breaks, collect all the mercury and store it in a labeled, sealed container. See Appendix 4B.
4. The heat of vaporization of water is 539 cal/g or 2255 joules (J) per gram. For ethyl alcohol, it is 205 cal/g (858 J/g). One calorie is equivalent to 4.184 joules.
5. Try equal masses of each liquid and other pure solvents and measure temperature over time. Plotting three curves on the same grid may give good comparative information. Plot time on the horizontal axis.
6. The students are examining differences due to the combined effects of the energy of vaporization and the rate of vaporization. Both result from differences in molecular structure of each type of molecule.

ANSWERS TO THE QUESTIONS

1. Probably the acetone.
2. Acetone requires less energy to leave the liquid; thus, it evaporates (or vaporizes) readily. However, it causes a large temperature drop because a lot is vaporizing in a short time.
3. Probably acetone, perfume or alcohol, and water. (Water requires so much energy that it cannot get it merely from the room-temperature surroundings in a short time.)

44. How Temperature Affects the Motion of Molecules

Molecules of a substance are in constant motion. In a solid, the motion is often restricted to vibration in a fixed position. In a liquid such as water, the molecules can move around more but are still somewhat restricted because of their attraction to other molecules. In a gas, this molecule-to-molecule attraction is overcome, and the molecules have unrestricted movement; thus, gases can fill whatever volume is available to them. When molecules are heated, they move faster and with more energy. In this activity we will use a drop of dye to estimate how temperature affects molecular motion.

Materials

1. Two tall glass containers.
2. Hot water.
3. Cold water.
4. Food coloring or black ink.

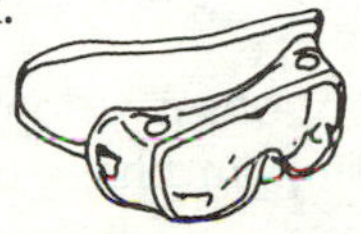

Procedure

1. Fill a large beaker or tall clear drinking glass about three-fourths full with very hot water from the hot water faucet.
2. Add the same amount of chilled water to a similar beaker or glass. Let both containers sit for a minute or so to minimize "thermal currents".
3. Add one drop of food coloring or ink to the center of each container.
4. Record your observations.

Reactions

The dye molecules have a tendency to separate as much as possible when the drop is placed in the water. In the hot water, the increase in temperature gives the dye molecules more energy and makes them separate faster; thus, the dye diffuses rapidly and evenly in the hot water. In the cold water, the process is much slower because the molecules have less energy. Temperature is actually a measure of movement (kinetic) energy.

Questions

1. Describe the fate of the dye in the hot water and in the cold water.
2. Does the dye in the cold water container drop to the bottom? Why? Does the dye in the hot water container drop to the bottom? Why?
3. What effect does temperature have on the motion of molecules?

Notes for the Teacher

BACKGROUND

This activity is a simple way to show that heat causes an increase in the motion of molecules. When a drop of dye is added to a container of hot water, the thermal currents and the motion of the energetic dye molecules cause a rapid dispersion of the dye, and a uniform solution results in just a few seconds. In a container of cold water in which the motion of the molecules is not as fast, the dye slowly mixes with the water, forming long streams. These streams fall to the bottom of the container before thorough mixing occurs.

The molecules of dye will try to separate as much as possible when the drop is placed in water. A chemist would say that a tendency toward maximum randomness of the dye molecules is exhibited. This condition is also called *entropy*.

MATERIALS

1. Hot water: Use hot water from the tap or heat some water. Do not use boiling water. You might want to heat a large container of water on a hot plate and dispense small amounts to each student pair. Do not place the dye in the water while it is being heated; thermal currents will cause the dye to mix quickly.
2. Cold water: Water from a drinking fountain works well. The water should be at least 10 °C below room temperature for best results.
3. Dye: Food coloring works well. Black ink works best and is more easily visible.

TEACHING TIPS

1. This activity is a good one for stressing the importance of observation. You might have the students do the activity without discussion and have them "discover" the effect of temperature.
2. The taller the glass container, the better the effect will be.
3. If students work in pairs, have them add the dye to the two containers at the same time.
4. Stress the point that the dye, because it is slightly more dense than water, tends to fall to the bottom because of gravity. The thermal density currents, the energy of the dye molecules, and the temperature gradients cause the swirling action of the dye as it dissolves in the water.
5. Emphasize that the dye in cold water will eventually become uniformly dispersed.
6. During the observation process, you could introduce the term "diffusion". Open a bottle of perfume or ammonia so that students can notice that gas molecules diffuse more rapidly than liquid molecules. There are not so many molecules to bump along the way.
7. Water molecules have been estimated to move faster than 1000 mph at room temperature. They move much faster when heated.

ANSWERS TO THE QUESTIONS

1. In the container of hot water, the dye swirls with more intensity than in the cold water and becomes thoroughly mixed within a few seconds. In the cold water, the dye forms long streamers and is not thoroughly mixed even after a minute or two.
2. The dye falls to the bottom in the cold water before it becomes mixed. It is mixed in the hot water before it reaches the bottom. This situation causes a clear area near the bottom of the hot water container.
3. Increasing the temperature increases the motion of the dye molecules and thus produces more mixing of the added dye and an increase in randomness of the dye molecules.

45. Temperature and Reaction Rates

For chemical reactions to occur, atoms, molecules, or ions must collide with each other. Usually, only a few collisions result in reactions that form new products because all conditions must be favorable when particles collide to form a new product. The orientation of colliding particles and the amount of kinetic energy must be just right, and the concentration of reacting particles must be sufficiently high for a reaction to occur. We can do several things to increase the chance of product formation when particles collide. In this activity we will vary the temperature of the reactants and see how it affects the formation of products.

Materials

1. Sodium thiosulfate ($Na_2S_2O_3 \cdot 5H_2O$) solution, 1%.
2. Vinegar (5% acetic acid solution).
3. Three large test tubes.
4. Three beakers, 250 mL or 400 mL.
5. Burner.
6. Tripod and wire gauze.
7. Clock or timing device.

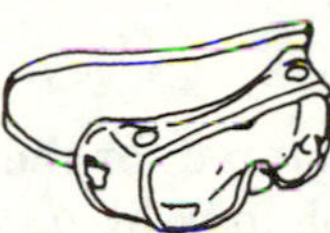

Procedure

1. Prepare a hot water bath by heating a beaker of water to 70–80 °C on a hot plate or on a tripod with a burner. Fill another beaker with ice to form an ice bath. Fill a third beaker three-fourths full with water at room temperature.
2. Place 10 mL of sodium thiosulfate solution in each of three test tubes. Put one in an ice bath, one in a beaker of boiling water, and the other in a beaker of water at room temperature. Leave the tubes in the beakers for 2–3 min.
3. With the help of another student, quickly pour 10 mL of vinegar into each of the tubes.
4. Stopper the tube and mix well. Record the time.
5. Notice the reaction that occurs. Observe and record which test tube reacted first, second, and third. Record the time when each tube reaches greatest cloudiness. Record the temperature of each of the three tubes.
6. Carefully smell each of the three tubes. Fan a breeze toward you over the mouth of the test tube to detect the odor. NEVER place your nose or face directly over any odorous chemical. Can you detect a similar odor in each tube?
7. Explain your observations on the basis of energy and motion of molecules, atoms, or ions.

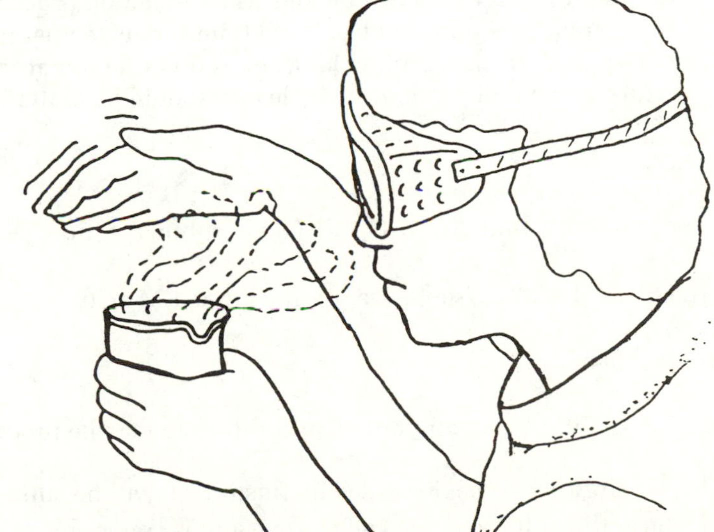

Reaction

Sodium thiosulfate ($Na_2S_2O_3 \cdot 5H_2O$) is very unstable in an acidic solution. When it reacts with acetic acid (vinegar, CH_3COOH), the thiosulfate ion decomposes to produce elemental sulfur, which gives the solution a cloudy appearance, and sulfur dioxide, which you might be able to detect because of its odor:

$$S_2O_3^{2-}(aq) + 2H^+(aq) \longrightarrow S(s) + SO_2(g) + H_2O(\ell)$$

In the cold tube, the energy of the ions is lowest, and few ions have the energy required to react. Therefore, the reaction and thus the formation of the precipitate of sulfur take longer to occur. In the hot beaker, more ions have the amount of energy required to react; the reaction is fastest and more sulfur molecules are formed in the shortest time.

Questions

1. What must molecules, atoms, or ions do so that a chemical reaction will take place?
2. What effect does temperature have on a chemical reaction?
3. How could you tell that a chemical reaction had taken place?
4. Make a graph of your data by plotting the time (y axis) required for the reaction to occur versus temperature (x axis). What does your graph look like?
5. From your graph, can you predict how long a reaction at 50 °C would take? Try it!

Notes for the Teacher

BACKGROUND

The kinetic molecular theory is one of the foundations upon which an understanding of chemistry is built. Molecules, atoms, and ions are constantly in motion. They must collide with sufficient energy for product formation to occur. In this activity students can slow down particles by decreasing temperature and speed up particles by increasing temperature. They will easily see the cloudy white precipitate and smell the odor of sulfur dioxide as the reaction proceeds. They can also check their predictive powers by plotting time versus temperature and estimating the time required for the reaction to occur at other temperatures. In this case, temperature is the independent variable and should be plotted on the x axis.

SOLUTIONS

1. Sodium thiosulfate ($Na_2S_2O_3 \cdot 5H_2O$) is a dilute solution. Dissolve 1.0 g in 100 mL of water.
2. Acetic acid is 5%. Use vinegar or any other 1 M acid.

TEACHING TIPS

SAFETY NOTE: A small amount of sulfur dioxide may be produced.

1. Each solution can be discarded by flushing down the sink.
2. Sodium thiosulfate is also called *hypo* and is commonly used in photography.

3. Sulfur molecules usually contain eight sulfur atoms (S_8). However, chemists most often use only an "S" to represent sulfur in a chemical equation.
4. Sulfur dioxide, when it becomes wet with water, forms sulfurous acid (H_2SO_3). Sulfur dioxide is produced by power plants. When sulfur dioxide mixes with atmospheric water, *acid rain* is formed. Have students research acid rain in the library.
5. Vinegar is 5% acetic acid. Try other acids; they will also work well.
6. You might try projecting these reactions in Petri dishes on the overhead projector. You might get a nice "sunset" effect on the screen. Use about 5 mL of each solution.
7. The sulfur that is produced is called *colloidal sulfur.* It is in a suspension because sulfur does not dissolve in water.
8. Sulfate is SO_4^{2-}. *Thio* means that the ion contains one less oxygen and one more sulfur atom, or $S_2O_3^{2-}$.

ANSWERS TO THE QUESTIONS

1. Reactions may occur when molecules, atoms, or ions collide.
2. An increase in temperature causes an increase in the average kinetic energy of the reacting species, which increases the rate of reaction.
3. A cloudy white precipitate formed, and a pungent, toxic gas was emitted.
4. When the three dots are connected with a smooth line, a curve should result.

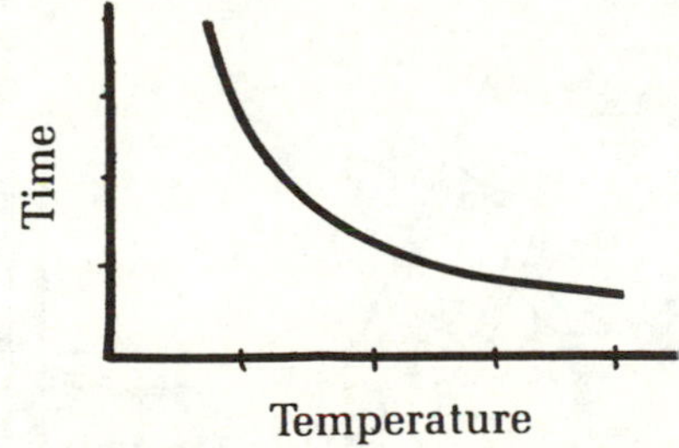

5. By picking any temperature on one axis and reading the corresponding time on the other axis, students can predict, with a fair amount of accuracy, the time required for the reaction to occur at any temperature on the graph.

46. Reaction Rate and Temperature of Vinegar

What will happen to the time it takes to react a piece of metal (magnesium) with vinegar if the vinegar is heated? What will you notice about the time if the vinegar is heated carefully and the times are studied at each temperature increase of 10 °C? You will study the reaction times and make a graph of the temperature on the *x* axis and the time on the *y* axis. The shorter the reaction time, the faster the reaction rate will be.

Materials

1. Two beakers for water bath, 250 or 400 mL.
2. Six test tubes to fit in the beaker.
3. Vinegar for each test tube.
4. Thermometer.
5. Magnesium ribbon, 12–15 cm.
6. Heat source (hot plate, Bunsen burner, alcohol burner).
7. Ice cubes.

Procedure

1. Place an equal volume of vinegar in each test tube.
2. Cut the magnesium ribbon into six pieces of equal length between 1 and 2 cm.
3. Place one test tube in a water bath in which you have placed a few ice cubes to make the water about 10 °C below room temperature.
4. After 2 min, when the temperature of the vinegar should be the same as that of the water bath, drop in a piece of magnesium ribbon. Record the time it takes for the ribbon to react completely with the acid in the vinegar.
5. Repeat with a test tube of vinegar that is at room temperature. What is room temperature? How do you know?
6. Meanwhile, begin heating water in another water bath. Place the remaining four test tubes of vinegar into the water bath. Put in the thermometer.

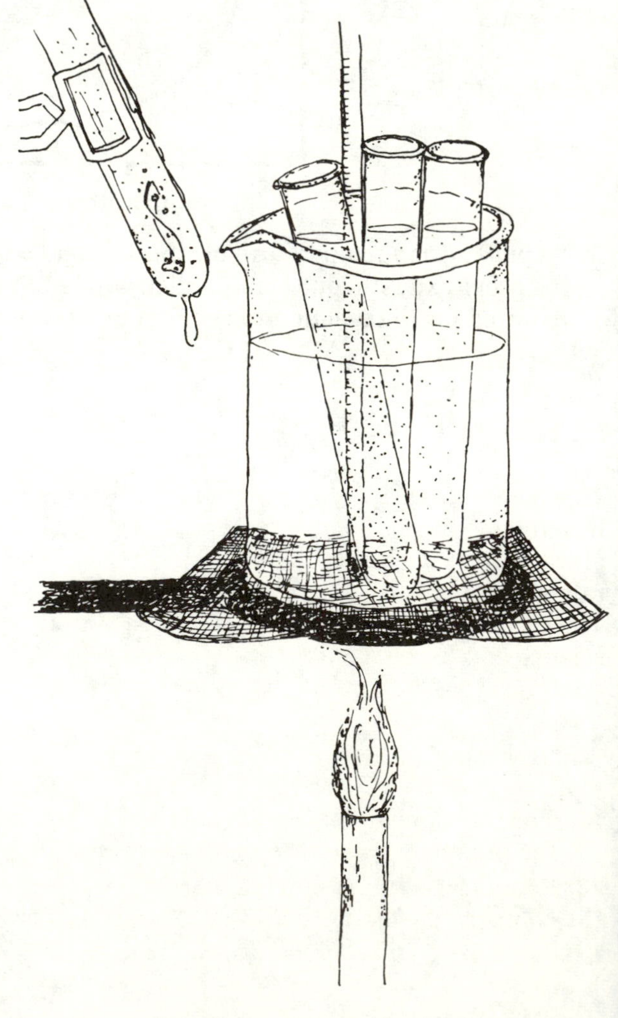

7. When the bath is 10 °C above room temperature, remove one of the tubes, record the time, and drop in a piece of magnesium ribbon. Observe the time it takes to react completely.
8. Keep an eye on the temperature of the water bath so that you can remove a tube each time the temperature goes up another 10 °C.
9. Continue to take reaction times at increments of 10 °C until you have six values. The lowest value should be at 10 °C BELOW room temperature, and the highest value should be at 40 °C ABOVE room temperature.
10. Make a graph to show the relationship between temperature increases of 10 °C and reaction time. Because the temperature of the reaction is controlled by you, it is the "independent variable" and should be placed on the horizontal axis, that is, the *x* axis.

Reaction

Generally, metals react with acids to produce hydrogen gas and the dissolved salt of the metal. Some metals, such as gold and silver, are so stable that they will not react under ordinary conditions. Some metals, such as sodium and calcium, are so reactive that they will react with water even if no acid is present. Magnesium reacts with the acid in vinegar, which is acetic acid:

$$Mg(s) + 2CH_3COOH(aq) \longrightarrow H_2(g) + Mg^{2+}(aq) + 2CH_3COO^-(aq)$$

Mg^{2+} is the symbol for the dissolved form of magnesium. The atoms of this form have lost two electrons each and are called *ions.* CH_3COO^- is called the *acetate* ion.

Questions

1. Why must all the volumes of vinegar and lengths of magnesium ribbon be the same with each trial?
2. Look at your results and state the relationship between a temperature change of 10 °C and the reaction rate.
3. What are the bubbles you see in the reaction?

Notes for the Teacher

BACKGROUND

This activity will impress upon students the great increase in reaction rate caused by a temperature increase. They will be surprised to find that an increase in temperature of 10 °C yields more than a small increase in rate. A general rule is that the rate will actually double, in most cases, when the temperature increases by 10 °C. The graphing portion of the activity allows students to practice an essential skill in science and to see the usefulness of creating a graph to find a mathematical relationship.

TEACHING TIPS

1. The reaction will be very slow in the colder tubes. You might set them up before the prelab discussion so that the "reaction time" is not "wasted time".
2. Room temperature is read on the thermometers. You might have the class compare the temperatures. The room-temperature readings often vary.

3. The reactions do not need to be done at precise intervals of 10 °C. When the graph is made, the real temperatures should be plotted. From the graph, the relationship between the temperature changes of 10 °C and the rate of reaction will be clear.
4. The temperature of the vinegar changes as the reaction proceeds because the tube was taken from the water bath. Students should notice this change. Ask for suggestions about how to avoid the irregularity, or ignore the irregularity because each trial will experience this same variable.
5. Teams of four students work well with this activity because of the many tubes and times to watch.
6. Be sure that the graphs are constructed correctly with equally valued intervals on the axes and that real times and temperatures are plotted.
7. You could repeat this experiment using dilution of the vinegar instead of temperature as the variable. Plot concentration versus time.

ANSWERS TO THE QUESTIONS

1. To control variables so you know that temperature is making the difference.
2. The reaction rate will double for each temperature increase of 10 °C.
3. Hydrogen gas.

Chemistry Around the House

47. Acids and Bases in the Bathroom and Kitchen

Many of the substances ordinarily found in the home are acids or bases. These include soaps, cleaning solutions, medicine chest items, and foods. In this activity we will first prepare some standard acid and base solutions to represent a range of pH values. The pH is a measure of acidity or basicity. We will use cabbage juice as a colored indicator of acidity or basicity. Then we can determine the pH values of various common substances by comparing them with these standards. The pH 7 is neutral. Less than pH 7 is acidic. A value greater than pH 7 is basic. **This activity will take two class periods.**

Materials

1. Red cabbage.
2. Lemon.
3. White vinegar.
4. Boric acid solution.
5. Sodium bicarbonate solution.
6. Borax solution.
7. Washing soda solution.
8. Sodium hydroxide solution.
9. Beakers, 100 and 250 mL.
10. Eight test tubes.
11. Burner.

Procedure

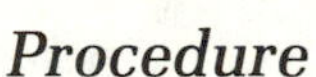

SAFETY NOTE: Wear safety goggles and plastic disposable gloves.

Part A: Preparing the pH Standards.

1. Number and label eight test tubes as follows: tube 1, pH 2; tube 2, pH 3; tube 3, pH 5; tube 4, pH 7; tube 5, pH 8; tube 6, pH 9; tube 7, pH 12; and tube 8, pH 14.
2. Prepare the indicator by shredding enough red cabbage to fill a 250-mL beaker about one-fourth full. Add about 100 mL of water and gently heat the cabbage until it boils. Continue to boil until the solution is deep purple. Allow the solution to cool. Pour off the liquid into a clean beaker. This liquid is the indicator solution.
3. Add 3 mL of indicator solution to each of the eight test tubes.
4. Add the following materials to the tubes:
   Tube 1: 10 mL of lemon juice (freshly squeezed, no seeds).
   Tube 2: 10 mL of white vinegar.
   Tube 3: 10 mL of boric acid solution. Avoid contact with the skin.
   Tube 4: 10 mL of water.
   Tube 5: 10 mL of sodium bicarbonate solution.
   Tube 6: 10 mL of borax solution. Avoid contact with the skin.
   Tube 7: 10 mL of washing soda solution.
   Tube 8: 10 mL of drain cleaner solution. Avoid contact with the skin.
5. Gently mix the solutions in each tube by tapping the tube or by stirring with a rod.
6. Note the colors. Do tubes 1–8 show a gradual change from red to yellow?

Part B: Determining the pH Values of Other Common Materials.

7. Make solutions of things around the house. Add indicator solution and determine the approximate pH by comparing the colors of the solutions with those of the standards.
8. You may want to try the following: shampoo, household ammonia, fruit juices, baking powder, antacid tablets, salt, garden fertilizer, and soil.
9. Record your observations.

Reaction

Red cabbage juice contains a substance called *anthocyanin*, which changes color when hydrogen ions (acid) are added or removed from the molecule. Acid contributes hydrogen ions.

Questions

1. Determining the pH values of substances such as grape juice, catsup, and egg yolk is difficult. Why?
2. We did not make standards for pH 4, 6, and 10. Can you suggest substances from your activity to serve as standards for these pH values?
3. How do your results compare with those of your classmates?

Notes for the Teacher

BACKGROUND

An *indicator* is a chemical compound that has a characteristic color at a certain pH. The color of an indicator in a basic solution is different from its color in an acidic solution. The natural indicator from red cabbage juice is red in very acidic solutions (pH 1–3), violet in slightly acidic solutions (pH 4–6), blue-green in slightly basic solutions (pH 8–10), and yellow in very basic solutions (pH $>$12). The chemical name of the indicator in red cabbage juice is *anthocyanin*. See Activity 33 for a study of natural indicators.

TEACHING TIPS

SAFETY NOTE: Sodium hydroxide (lye) is caustic. Wear gloves, goggles, and a face shield when you mix this solution. Use care when mixing solutions of borax and boric acid.

1. Prepare the pH standard solutions as follows:
   A. Several lemons for squeezing.
   B. White vinegar: 10 mL per student or pair.
   C. Boric acid solution: Dissolve 2 g of boric acid in 200 mL of water.
   D. Sodium bicarbonate solution: Dissolve 2 g of sodium bicarbonate (baking soda) in 200 mL of water.
   E. Borax solution: Dissolve 2 g of borax (sodium borate) in 200 mL of water.
   F. Washing soda solution: Dissolve 2 g of sodium carbonate in 200 mL of water.
   G. Sodium hydroxide solution: Dissolve 2 g of solid drain cleaner (e.g., Drano) in 200 mL of water. BE CAREFUL! This solution gets hot. Do not get any on your skin or clothing. Use with caution.
2. If you have a large class, you might prefer to prepare a large beaker of red cabbage juice for everyone to use.

3. The standards should show a gradual change from red to yellow. Students may want to experiment with color change when an acidic solution and a basic solution react. Be careful if an acid is added to tubes 5 or 7 because carbon dioxide gas will form and cause frothing.
4. You might prepare some unknowns for the students to determine the pH values. Put a chart on the blackboard for them to record their data for class comparison.
5. Sodium bicarbonate is $NaHCO_3$. Sodium carbonate is Na_2CO_3. Borax is $Na_2[B_4O_5(OH)_4]\cdot 8H_2O$. Drain cleaner is mostly sodium hydroxide ($NaOH$). Vinegar is a 5% solution of acetic acid (CH_3COOH).
6. Several other natural indicators can be used for this activity: beet juice, carrot juice, grape juice, blueberry juice, and flowers including the blue iris, purple dahlia, and purple hollyhock. See Activity 33.

ANSWERS TO THE QUESTIONS

1. Substances with colors that interfere with the color of the indicator cannot be determined by this method.
2. Answers will vary. Encourage students to develop their own standards.
3. Answers will vary, but the results should be in fairly close agreement.

48. Chemist in the Kitchen: Mystery Powders

This experiment will let you learn about a few *physical* and *chemical* properties of several common substances ordinarily found in the home. You will be given four powders. We will call these A, B, C, and D. You will also have three different liquids (I, II, and III). The object of this activity is to find some properties of the powders by using the liquids. When you can tell the powders apart by their properties, you will be given a mystery powder that will contain A, B, C, or D or any combination. Using the liquids and your other observations, you will identify your mystery powder.

Materials

1. Powders (A, B, C, and D).
2. Magnifying glass or binocular microscope.
3. Liquids (I, II, and III).
4. Plastic sheet or glass plate.

Procedure

1. Prepare two grids by dividing a 4- × 4-in. piece of paper into 12 equal squares as shown. Cover one grid with the acetate or plastic so that you can keep track of your observations.
2. Place pea-size samples of each powder (A, B, C, and D) in the proper squares across from each letter.
3. Carefully examine each powder for its *physical* properties. Record your observations on the similar grid.
4. Place 1 drop of each liquid (I, II, and III) on each solid in the squares so that you have all combinations of liquids and solids to observe. Record the *chemical* properties.

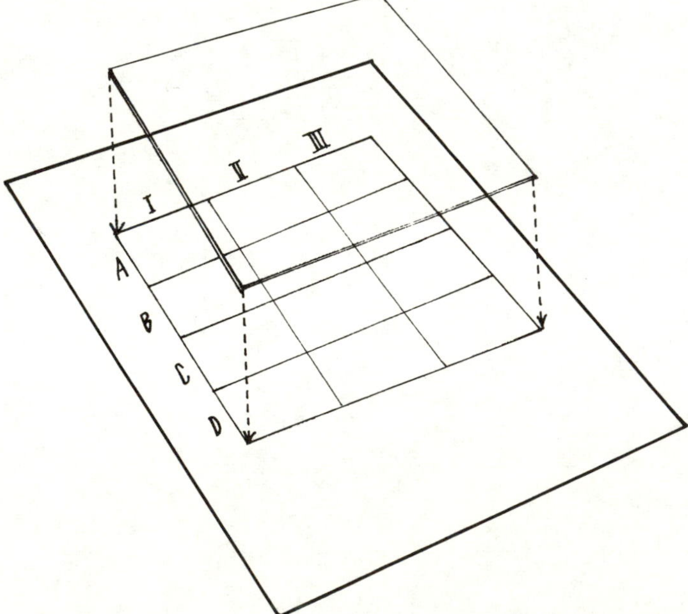

5. Obtain an unknown sample from your teacher. Using the procedure just given and your observations of A, B, C, and D with liquids I, II, and III, identify the components of the unknown.

Reaction

Physical properties can be studied by observing solubility, color, crystal structure, malleability (brittleness), and melting point.

Chemical properties can be studied by reacting the substance with another substance and observing the changes that occur, if any. Such changes may involve gas formation, color change, or precipitate formation. Chemical bonds are broken and new bonds between atoms are formed. We looked for chemical properties when we added the liquids to the solids.

Questions

1. Write a paragraph describing how you were able to identify your mystery powder.
2. Your observations in this experiment are limited to the senses. How could you extend your ability to make observations?

Notes for the Teacher

BACKGROUND

This activity is a good experiment with which to begin a laboratory program in chemistry. The emphasis here is on the process of reasoning from laboratory observations. The intent is not to identify the powders and liquids chemically, but merely to observe how they react, to organize the observations, and to make predictions on which to base the eventual identification of mystery powders. The students will be detectives as they work. They will also be experiencing the experimental nature of chemistry. Because of the reasoning experience provided by this activity, it is one of the most important chemistry experiments a student can do. Encourage interaction among students.

TEACHING TIPS

SAFETY NOTE: Keep close track of all cups and spoons so that they are not mistaken for food implements.

1. The powders recommended are the following:
   A. Baking soda (sodium bicarbonate, $NaHCO_3$).
   B. Laundry starch $(C_6H_{10}O_5)_n$.
   C. Baking powder (a mixture of baking soda and a solid acid).
   D. Salt (sodium chloride, NaCl).
2. The liquids should be the following:
   I. Vinegar (acetic acid, CH_3COOH).
   II. Iodine solution [iodine crystals and potassium iodide (KI)]. Make the solution dilute by adding 2.5 g of KI to 500 mL of water. Add 1 g of iodine crystals, or add tincture of iodine to the water to make the solution yellow.
   III. Distilled water.
3. Other powders can be added to increase the challenge. Suggestions are sugar, plaster of Paris ($CaSO_4$), cornstarch, and nonfat dry milk. Other tests may be suggested, for example, combustibility.
4. Place the powders in paper cups and the liquids in dropper bottles.
5. Glass plates and plastic for an overhead projector work well to cover the grid.

6. You may want to reproduce grids for the students, including one on which they can make their observations.
7. Do not allow students to taste these chemicals.
8. Prepare the mystery powders in the following combinations as well as singly:

1-AB	5-CD
2-AD	6-ABD
3-BC	7-BCD
4-BD	

ANSWERS TO THE QUESTIONS

1. Answers will vary. Look for the ability of the student to organize thoughts and express observations rather than inferences or interpretations.
2. Use equipment such as thermometers, balances, and microscopes.

49. A Chemical Weather Predictor

Some chemicals change color when water is added to them. Some colored chemicals owe their color to the water molecules that are always associated with them (unless they are heated). Cobalt chloride is an interesting salt because when it is dry on paper with table salt in it, the cobalt is one color. When damp, it is another color. Can you find out what those colors are?

Materials

1. Cobalt chloride solution.
2. Sodium chloride solution.
3. Filter paper.
4. Small beaker or plastic cup.

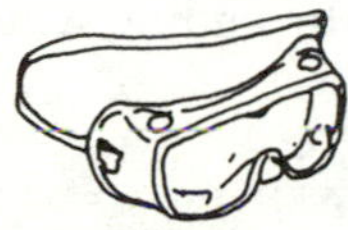

Procedure

SAFETY NOTE: Wear safety goggles and plastic disposable gloves.

1. Mix equal volumes of the two solutions in a small beaker or cup. Avoid contact with the cobalt chloride solution and filter papers. Cobalt chloride is toxic.
2. Dip half of the filter paper into the solution, using tongs.
3. Allow the paper to dry. Wrap it around a beaker or test tube of hot water to hasten drying.
4. Observe differences in color between the wet and the dry cobalt chloride paper.
5. Place the dry cobalt chloride paper near an open window on a humid day and see what happens.

Reaction

Cobalt atoms in salts are positive ions with a 2+ charge. They attract negative particles such as chloride ions (Cl^-) and the oxygen end of water molecules. When most of the negative species around the cobalt ion are water molecules, the ion absorbs light so that it appears pink. When the paper is dry, the water molecules are gone (evaporated). The negative chloride sticks to the positive cobalt ions, and the cobalt appears blue. The water and the chloride are different and cause the electrons in the cobalt to absorb different energies. These different energies result in different colors absorbed by the cobalt.

$$\underset{\text{pink}}{2Co(H_2O)_6^{2+}(aq)} + 4Cl^-(aq) \longrightarrow \underset{\text{blue}}{Co(CoCl_4)(s)} + 12H_2O(\ell)$$

Questions

1. What other purpose might the sodium chloride have in addition to supplying more chloride ions?
2. How many times can you cycle the cobalt chloride paper between colors?
3. Suggest a practical application for the cobalt chloride paper.

Notes for the Teacher

BACKGROUND

Many students have seen the blue weather papers made of absorbent material. They turn pink when the weather is humid and are, therefore, predictors of wet weather to come. This change of the species surrounding the cobalt ions in the paper causes the change in color (see student page) and can lead to some discussion of sophisticated concepts of electronic structure.

SOLUTIONS

Cobalt chloride solution is made by adding water to 12 g of $CoCl_2 \cdot 6H_2O$ to make 100 mL of solution. Sodium chloride solution is made by adding water to 6 g of NaCl to make 100 mL of solution.

TEACHING TIPS

1. A small amount of solution will go a long way, and many students can share the same beakers.
2. The concentrations of the solutions are not too important as long as the pink color is noticeable on the paper.
3. A hair dryer works well for drying the papers.
4. Have students place the blue paper in several places around the school. Monitoring humidities is fun and will interest other students in the school.

ANSWERS TO THE QUESTIONS

1. The sodium chloride picks up moisture from the air.
2. Indefinitely, as long as water does not wash the salts off the paper.
3. It can be used to monitor humidity or identify the presence of water anywhere. It can also be used to detect water in gasoline samples and in certain chemicals.

50. Making Glue

White glue is often made from the protein in milk called *casein*. The casein is separated from milk by processes, which you will do here, called *coagulation* and *precipitation*. At the factory, the casein is dried and ground up before it is made into glue. Casein is also used in some paints and to make a type of plasticlike button. You will make glue using milk.

Materials

1. Skim milk, 125 mL.
2. Beaker, 250 mL.
3. Vinegar (acetic acid, CH_3COOH), 25 mL.
4. Heat source.
5. Sodium bicarbonate (baking soda, $NaHCO_3$), 1 g.
6. Funnel and filter paper.
7. Stirring rod.
8. Water, 30 mL.
9. Graduated cylinder, 125 mL.

Procedure

1. Place about 125 mL of skim milk in a 250-mL beaker.
2. Add about 25 mL of vinegar, which is an acidic solution.
3. Set up a heat source. Gently heat the milk and stir constantly until small lumps begin to form.
4. Remove the beaker from the heat and continue to stir until no more lumps form.
5. Allow the lumps (curds) to settle.
6. Filter the curds from the liquid (whey) by using filter paper and a funnel.

7. Gently press the filter paper around the curds to squeeze out the excess liquid through the filter paper.
8. Return the solid material to the empty beaker.
9. Add about 30 mL of water to the solid and stir.
10. Add about ½ tsp of sodium bicarbonate to neutralize any acid remaining from the vinegar. Watch for bubbles of gas to appear. Add a little more sodium bicarbonate until no more bubbles appear.
11. The substance in the beaker is glue. Test the adhesive properties of your product with various materials.

Reaction

Coagulating the protein and neutralizing the vinegar are chemical changes.

1. Proteins are sensitive to changes in acidity and heat. A coagulated protein, casein, is produced in this experiment by adding both acid (vinegar is 5% acetic acid) and gentle heat. The solid casein is a *precipitate.*
2. The excess acid from the vinegar is neutralized by the base, sodium bicarbonate. The gas, carbon dioxide, can be seen as bubbles:

$$\underset{\text{acetic acid}}{CH_3COOH(aq)} + NaHCO_3(s) \longrightarrow \underset{\text{sodium acetate}}{NaCH_3COO(aq)} + H_2O(\ell) + CO_2(g)$$

Questions

1. What is the purpose of the vinegar in making glue? What is the purpose of the heat?
2. How is the casein separated from the milk?
3. What is the purpose of the sodium bicarbonate (baking soda)?

Notes for the Teacher

BACKGROUND

This activity allows students to make and test a product that is much like casein glue, a commercial product. Students will have the opportunity to study a protein (casein in milk) and the reactions that this protein undergoes. Casein is the primary protein in cow's milk. In this activity students will observe the effects of heat and acid on a protein.

TEACHING TIPS

1. The reactions to make glue take place right away. The students will be able to enjoy the quick reactions and will also have the opportunity to practice laboratory skills.
2. In the presence of acid, the protein, casein, will clump. The molecules no longer repel each other but have an attraction for each other and thus form large particles, the curds.
3. Comparing the adhesion of this glue on different materials is interesting. Test to see if the glue is waterproof (it is).

4. See other milk investigations in this volume—Activities 59, 79, 84, and 100.

ANSWERS TO THE QUESTIONS

1. The vinegar acts to change the protein molecules so that they precipitate. The heat also causes the protein molecules to change.
2. The casein is separated by filtering it out of the liquid (whey).
3. The sodium bicarbonate neutralizes the acid.

51. Making Cleansing Cream

In this activity you will make cleansing cream that can be used to remove substances from the surface of skin and act as a moisturizer as well. Cleansing cream is an oil and wax emulsion in a large amount of water. The oil and wax are mixed in the water in tiny droplets about 1/1000 of 1 mm in diameter. When the cream is placed on the skin, the emulsion mixture separates, and dirt, grease, oil, and makeup are dissolved by either the oil or the water. All of the matter is lifted off when the cream is rinsed or wiped from the skin.

Materials

1. Paraffin wax cut into small pieces, 20 g.
2. Mineral oil, 125 mL.
3. Borax, sodium borate, 1.5 g.
4. Stearic acid, 1.5 g.
5. Fragrance.
6. Beaker or empty can, 400 mL.
7. Thermometer.
8. Balance.
9. Heat source.
10. Graduated cylinder, 100 mL.
11. Stirring sticks.
12. Small jars to hold cream.

Procedure

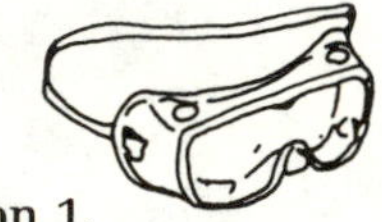

1. Prepare solution 1.
   A. Obtain 20 g of paraffin. Place it in the can or beaker.
   B. Label the can or beaker with your name and place it in a water bath so that the paraffin will melt. The water bath should be kept as close to 70 °C as possible.

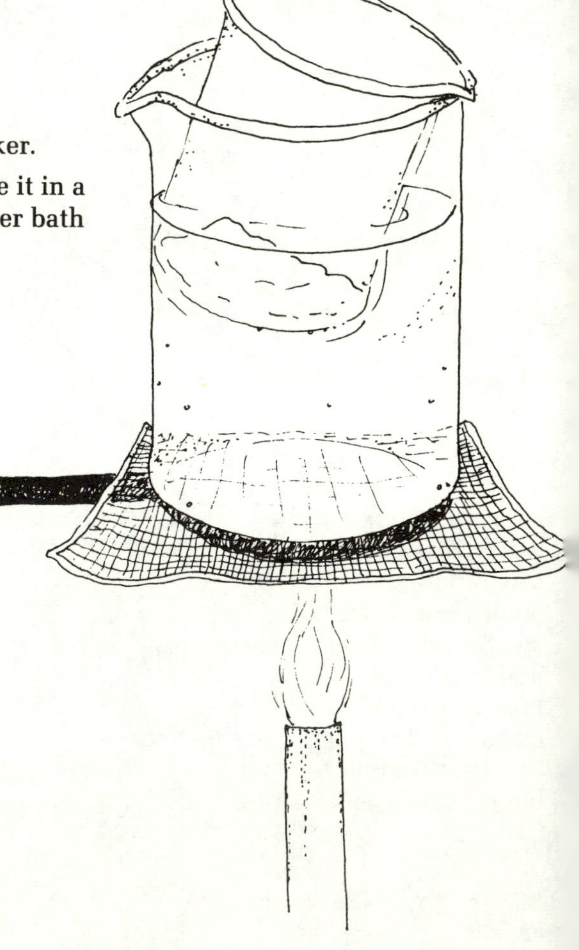

C. While waiting for the paraffin to melt, weigh out 1.5 g of stearic acid.
D. Measure 125 mL of mineral oil in a graduated cylinder.
E. When the paraffin is melted, add the mineral oil and the stearic acid.
F. Place the can in the water bath so that the contents will warm to 70 °C again.

2. Prepare solution 2.
   A. Heat 70 mL of water to 70 °C.
   B. Weigh out 1.5 g of sodium borate.
   C. Add the sodium borate to the warm water and stir until it all dissolves.
   D. Cool to 60 °C.

3. Make the emulsion. Here is where the skill is needed!
   A. While stirring constantly, slowly add solution 2 at 60 °C to solution 1 at 60–70 °C. If you wish to add perfume, add 1–5 mL at this time.
   B. Continue to stir until the cleansing cream cools but is still a liquid.
   C. Pour the cleansing cream into clean jars.
   D. Place all glassware in soapy water.

   Congratulations! You have now experimented like a chemist in the cosmetic business!

Reaction

The changes are all physical changes called *emulsification*. Emulsification occurs when a liquid other than water is mixed into (dispersed in) water in very small droplets. Imagine how small 1/1000 of a millimeter is.

Questions

1. Describe the product you prepared.
2. How small are the droplets of oil and wax that you mixed into water to make this emulsion?
3. How does this cream clean the skin?
4. How does this cream compare with commercial cream?
5. What are some other examples of emulsions?

Notes for the Teacher

BACKGROUND

An *emulsion* is a mixture that is formed when a nonaqueous liquid is dispersed in water. An emulsifying agent is required to keep the two liquids together and prevent separation. In this procedure, stearic acid, a common ingredient in soaps, is the emulsifying agent. Lecithin, an emulsifying agent found in egg yolks, also contains stearic acid as part of its molecule. In mayonnaise, olive oil is dispersed in vinegar, and lecithin is the emulsifying agent. Two other common emulsifying agents are bile, which allows fats to be digested, and soap, which disperses fats and oils during cleaning.

Paraffin and mineral oil are products of the distillation of crude oil. Sodium borate, also known as borax, helps to give the product cleansing properties.

TEACHING TIPS

SAFETY NOTE: Paraffin wax and mineral oil are combustible. Extinguish all flames in the room. Handle borax with care. Although sodium borate is used in many leading commercial cosmetic preparations, some individuals may be sensitive to it.

1. This activity is fun. It allows students to develop good laboratory techniques and helps them to develop confidence in the laboratory. At the same time, students learn firsthand about the activities of the cosmetic industry. They also have the experience of observing physical changes in the production of an emulsion. You may want to make the emulsion "mayonnaise". (See Activity 101.)
2. Prepare a water bath to keep at 70 °C for three to five teams of students. Use a pan of water and a hot plate or other heat source. Because paraffin and mineral oil are flammable, using a hot plate would be safest.
3. Using a disposable can for solution 1 will eliminate some cleanup after the activity.
4. None of the chemicals are toxic or harmful. All can be purchased in grocery stores or drugstores.
5. The temperature of the heated water and wax is kept low to prevent burns. Students should be careful with hot wax.
6. Have students bring their favorite fragrances from home. Perfumes, rose water, menthol, oil of peppermint, and oil of cinnamon may result in products that are much like some of the popular commercial creams.
7. If the emulsion is not stirred adequately, a water layer will form later in the bottom of the jar.
8. This cleansing cream is almost identical to the commercial product. However, it is good practice to discourage students from applying laboratory-produced substances to their skin.

ANSWERS TO THE QUESTIONS

1. Your product should be similar to commercial cream.
2. The droplets are too small to be seen.
3. The cream cleans because the oil part will dissolve some of the grease and dirt and the water part will dissolve other kinds of dirt. All of the matter is lifted off when the cream is removed.
4. Have a commercial variety of cleansing cream available for comparison.
5. Mayonnaise and homogenized milk are emulsions. Have students look up emulsions in the library.

52. Preparing Soap

People have been using soap to clean their clothes and themselves for a relatively short time. As recently as 1850, bathing regularly was not a common practice. Homemade soaps were harsh, and manufactured commercial soaps were expensive, costing five times the price of a quart of milk. Early soaps were produced by using the same procedure that you will use. The source of the hydroxide was a solution from soaking ashes in water. The source of the fat was lard from animals and general kitchen fat. Finer soaps can be prepared by varying the kinds of fat. **This activity will take 1–3 days.**

Materials

1. Fat, oil, shortening, or lard.
2. Sodium hydroxide (NaOH), 6 M.
3. Ethyl alcohol.
4. Two beakers, 100 mL.
5. Thermometer.
6. Heat source.
7. Wooden or plastic spoon or stick.

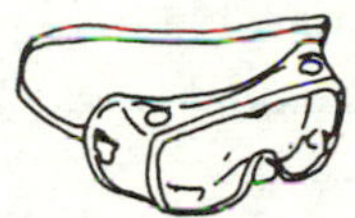

Procedure

1. Prepare 20 g (about 20 mL) of the fat mixture of your choice by melting any solid fat.
2. Allow the mixture to cool to about 45 °C.
3. Heat a mixture of 10 mL of 6 M sodium hydroxide and 10 mL of ethyl alcohol to about 35 °C. USE CARE: Sodium hydroxide is corrosive. Use gloves, goggles, and a face shield. Alcohol is flammable. Extinguish all flames in the room.
4. When the ingredients are at the desired temperatures, slowly pour the sodium hydroxide solution into the melted fats while stirring constantly with a wooden or plastic spoon or stick.
5. Stir until the saponification reaction is complete; a thick substance in which the stirrer will stand upright should be produced. If the ingredients were at the right temperature, this reaction should take 5–10 min.
6. If you wish to add scent to your soap, add about 1 mL of any essential oil, ground cloves, or cinnamon at this time.
7. Pour the soap slowly and evenly into a paper cup, which will serve as a mold.
8. Test the sudsing action of your soap by washing out the beaker.
9. Allow the soap to set for 1–3 days.
10. Using pH paper, test the alkalinity of your soap. Any number above 7, read from the color guide, is alkaline.

Reaction

A *fat* is a large molecule that breaks into four smaller ones when stirred with sodium hydroxide. Three of the new molecules are soap molecules, and one is glycerol, which keeps the soap moist. This reaction is called *saponification*.

$$
\begin{array}{lcll}
\overset{\displaystyle O}{\overset{\|}{R{-}C}}{-}O{-}CH_2 & & \overset{\displaystyle O}{\overset{\|}{R{-}C}}{-}O^{-}Na^{+} & HO{-}CH_2 \\
\overset{\displaystyle O}{\overset{\|}{R'{-}C}}{-}O{-}CH & + \; 3NaOH \longrightarrow & \overset{\displaystyle O}{\overset{\|}{R'{-}C}}{-}O^{-}Na^{+} & + \; HO{-}CH \\
\overset{\displaystyle O}{\overset{\|}{R''{-}C}}{-}O{-}CH_2 & & \overset{\displaystyle O}{\overset{\|}{R''{-}C}}{-}O^{-}Na^{+} & HO{-}CH_2 \\
\text{fat} & & \text{three soap molecules} & \text{glycerol}
\end{array}
$$

(R, R′, and R″ represent long-chain hydrocarbons.)

Questions

1. Describe the process of the saponification reaction.
2. How does sodium hydroxide change the fat molecule to form soap?
3. How do you think drain cleaner (sodium hydroxide) works to remove grease from the drain?
4. If you used pH paper, what was the pH of your soap? Is it alkaline, acidic, or neutral?

Notes for the Teacher

BACKGROUND

Making soap with the desired consistency is an art that is dependent upon the temperatures of the ingredients, the stirring time, and the quality of the fats. For one fat molecule, the products are always three soap molecules and a glycerol molecule. Many formulations and processes can be used for making soap. The field is worth extra study by interested students. Saponification was not identified as a predictable chemical process between sodium hydroxide and fat to produce soap until 1823.

SOLUTIONS

Prepare 6 M NaOH by adding 240 g of NaOH to enough water to make 1 L of solution. Use care. Wear gloves, goggles, and a face shield. Use a large beaker and stir constantly.

TEACHING TIPS

SAFETY NOTE: Ethyl alcohol is flammable. Sodium hydroxide is caustic. Use care with these chemicals. Extinguish all flames in the room.

1. See Activity 53 for a procedure to produce a clear soap from pure sodium stearate soap flakes.
2. If the soap does not form, the solution may be heated for 20 min in a beaker that is set in another beaker of boiling water and then poured into a 25% solution of sodium chloride. The hardened soap will float to the top of the solution.
3. Ethyl alcohol increases the solubility of the fat in the sodium hydroxide solution.

4. A mixture of solid fat with coconut or olive oil makes a fine-quality soap.
5. Constant stirring is necessary to keep the oil in contact with the sodium hydroxide.
6. While it cures, soap may be allowed to sit on a cheesecloth that is covering chopped pine needles or herbs.
7. Students should not use this soap if the pH is greater than 8.5 because it will probably be too impure and too alkaline. This procedure, however, is used by specialty soap makers.
8. "pH paper" can be made from simmered blueberries or purple cabbage. Alkaline substances turn the juice green. Acid turns it red.

ANSWERS TO THE QUESTIONS

1. Two clear liquids turn milky and thicken when mixed and stirred together for several minutes.
2. The bonds are broken in the fat and replaced by sodium atoms (ions). (See the Reaction section.)
3. The drain cleaner (NaOH) reacts with the grease in the drain to make soap in the drain.
4. The pH of the soap may be between 9 and 10 (alkaline). The pH of skin is slightly acidic; it is close to 6.

53. High-Quality Facial Soap

Pure soap can be dissolved in ethyl alcohol, glycerin, and water to produce the prized transparent soaps. Each batch can be custom-made with added fragrance, color, and extracts of herbs and oils. The shape of the soap is also customized. Glycerin soaps are good for the skin.

Materials

1. Pure soap flakes, 20 g (Ivory soap works well).
2. Ethyl alcohol, 15 mL.
3. Glycerin, 25 mL.
4. Water, 15 mL.
5. Beaker, 250 mL.
6. Hot plate.
7. Molds (gelatin molds are nice, and plastic Easter eggs are a good size).
8. Fragrances and food coloring (oil of cloves is traditional).

Procedure

BE CAREFUL !!!

1. Add soap, alcohol, glycerin, and water to the beaker. Heat at low setting until the solution is clear and the volume is reduced to between two-thirds and one-half of the original volume. Do not stir toward the end of heating. CAUTION: Alcohol is flammable. Extinguish all flames in the room.
2. Add color and fragrances (about 10 drops).
3. Prepare the molds by wiping them with glycerin. Pour the hot soap in the molds. Cool to harden.

Reaction

The dissolving of the soap molecules with the polar water, glycerin, and alcohol causes the light to pass through the solution rather than to be diffracted by soap pieces. The glycerin retained on the skin attracts moisture. Soap molecules are chains of 16–20 carbon atoms; one end of each chain dissolves well in water. Soap cleanses because the long chains of carbon atoms dissolve in the grease and dirt, and the end that dissolves in water causes the dirt and grease to wash away during rinses.

Questions

1. How could you tell the substances were making a solution?
2. Why does the soap need to “cook”?
3. Find out how soap works to clean grease and dirt.

Notes for the Teacher

BACKGROUND

This activity illustrates the principles of solutions, polar and nonpolar molecules, and the cleansing action of soap. Students will find it satisfying to synthesize something useful that can be customized.

TEACHING TIPS

1. Do not be alarmed if the solution is too thick to stir at the outset.
2. If kept simmering at a low setting, the soap should be ready to pour in about 45 min to 1 h. This experiment is a good after-school activity. Students will flock in to make gifts for their friends and families.
3. A substance that promotes moisture retention is called a *humectant*. Glycerin is also called glycerol.
4. Structural formulas that you might find useful. The lines projecting from C, carbon, represent hydrogen atoms.

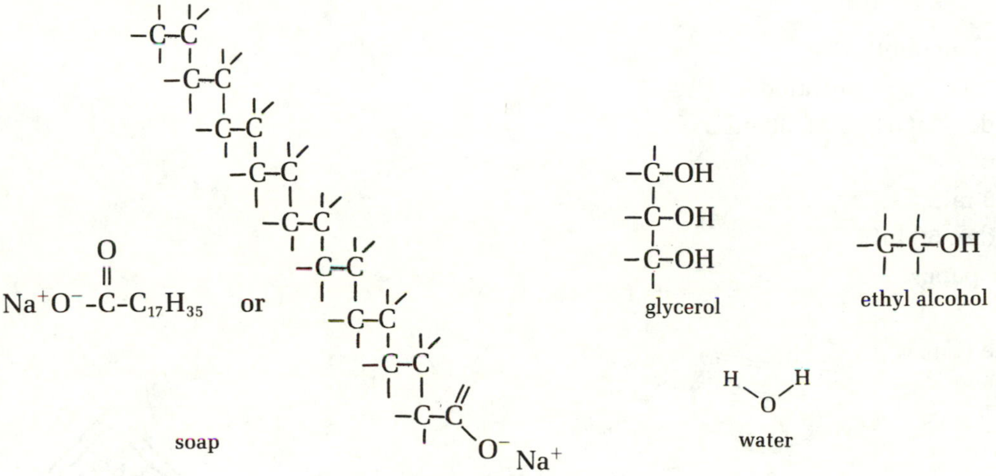

$Na^+O^- - \overset{\overset{O}{\|}}{C} - C_{17}H_{35}$ or $-C-C-C-C-C-C-C-C-C-C-C-C-C-C-C-C-C(=O)O^- \; Na^+$

soap

$-C-OH$
$-C-OH$
$-C-OH$

glycerol

$-C-C-OH$

ethyl alcohol

$H-O-H$

water

ANSWERS TO THE QUESTIONS

1. The individual properties of the substances could not be seen; also the system became clear.
2. To reduce the liquid in the solution so that the soap will be solid when cool.
3. See any college general chemistry book. Many high school books also have a reference to the mechanism of soap action. Consult the encyclopedia.

54. Preparing a Detergent

Detergents are used more than soaps in washing clothes and dishes because detergents work in hard water. Hard water has many minerals dissolved in it. These minerals react with soap molecules and form soap scum, which will not dissolve and will not be washed away. However, detergent molecules can do their job of soaking up grease and dirt even if minerals are in the water. You will make detergent from sodium hydroxide and a special alcohol. You may have completed Activity 52, in which you made a soap from sodium hydroxide and a fat. You can compare these reactions and the cleaning action of detergent and soap.

Materials

1. Dodecanol (lauryl alcohol), 15 g.
2. Sulfuric acid (H_2SO_4), concentrated.
3. Sodium hydroxide (NaOH), 6 M, 30 mL.
4. Phenolphthalein.
5. Two beakers, 250 mL.
6. Graduated cylinder, 50 mL.
7. Funnel and filter paper.
8. Test tubes.
9. Calcium chloride ($CaCl_2$).

Procedure

CAUTION: You must be very careful when pouring these solutions. If any solution spills on your skin or clothing, wash immediately! Wear gloves, goggles, and a face shield.

1. In two beakers, A and B, place the following substances:
   Beaker A: 15 g of dodecanol and 10 mL of sulfuric acid, H_2SO_4.
   Beaker B: 30 mL of 6 M NaOH and 3 drops of phenolphthalein.
2. While stirring constantly, SLOWLY add the acidic solution in beaker A to the basic solution in beaker B until the pink phenolphthalein just turns colorless.
3. Filter the solution and let the precipitate dry on the filter paper. You have produced a *syndet*, or rather, a synthetic detergent.
4. Test the detergent for its ability to produce suds in soft or distilled water. Compare it with soap.
5. Make some hard water by dissolving some calcium chloride in water.
6. Test the detergent and soap in two test tubes of hard water.

Reaction

Dodecanol (lauryl alcohol) is used in many commercial detergent molecules. The reaction in this activity is less complex than the series of reactions used for making most detergents.

$$CH_3CH_2CH_2CH_2CH_2CH_2CH_2CH_2CH_2CH_2CH_2CH_2OH$$

dodecanol

$$CH_3CH_2CH_2CH_2CH_2CH_2CH_2CH_2CH_2CH_2CH_2CH_2\text{-}\overset{\overset{\displaystyle O}{\|}}{\underset{\underset{\displaystyle O}{\|}}{S}}\text{-}O^-Na^+$$

sodium dodecylsulfonate (also called sodium lauryl sulfate)

When phenolphthalein is added to a base such as NaOH, it becomes bright pink. As the base is neutralized by the acid, the phenolphthalein also reacts. It is colorless in neutral or acidic solutions.

This detergent is called sodium dodecylsulfonate and is a surfactant. A surfactant acts on water and grease at the surface so that dirt dissolved in either the water or the grease can be washed away. A surfactant also lowers the surface tension of water. See Activity 11 for more on surface tension.

Questions

1. Describe the sample of dodecanol. *Dodec* means twelve in Latin. Why is this name used for this molecule?
2. Examine labels of shampoos and cleaning compounds. How many can you find that contain sodium dodecylsulfonate (or sodium lauryl sulfate)?
3. What happens to the phenolphthalein during the reaction to make detergent? Discuss the acidity or basicity of each solution in this experiment.
4. Compare the cleansing action of soap with that of detergent in each of the tests you performed.

Notes for the Teacher

BACKGROUND

Detergents are good cleaners because the long chain of carbon groups (CH_2) can dissolve in any oil and grease in which dirt is trapped. The other end of the molecule dissolves in water and allows the dirty spots to be rinsed away. Soap works the same way. However, soap molecules react with any calcium or magnesium ions in the water and cause an insoluble scum. Detergent molecules do not make scum. The American Chemical Society estimates that the amount of detergent products for home cleaning and personal care will total 15 billion lb in 1995. In addition to surfactants such as the one produced in this activity, commercial detergents contain whiteners, bleaches, enzymes, fragrance, and sudsing agents. Although the amount of suds often has nothing to do with the cleansing power of a soap or detergent, consumers have been convinced by advertising that it does.

SOLUTIONS

The sulfuric acid is concentrated, 18 M. To prepare 6 M NaOH, add 240 g of NaOH to enough water to make 1 L of solution. Use care. Use a large beaker and stir constantly. Wear gloves, goggles, and a face shield.

TEACHING TIPS

SAFETY NOTE: Use care when handling the acids and bases in this activity.

1. The teacher may want to dispense the 10 mL of concentrated sulfuric acid from the vented hood or other controlled area. Dispensing from a buret involves a minimal amount of handling and dripping.
2. If you are concerned about students handling concentrated acids, consider this activity as a demonstration through Procedure 2. If the amounts are increased, each student can have a sample of the suspension produced in Procedure 2. Students could then begin with Procedure 3.
3. Discussing the role of the neutralization reaction in step 2 is useful. Once the phenolphthalein is colorless, no additional acidic solution should be added.

4. A nice way to dry filter papers is to spread them on the type of hot tray used to keep dishes warm during buffet serving.
5. If your students are interested, the structures of soap and detergent molecules can be compared.

ANSWERS TO THE QUESTIONS

1. The dodecanol is a white solid. Each molecule contains 12 carbon atoms.
2. Many shampoos contain the surfactant that was synthesized in this activity.
3. As the acid is added to the base, the phenolphthalein reacts to change from pink to colorless. Phenolphthalein is pink in the base, sodium hydroxide, and colorless in solutions that are neutral or acidic.
4. Soap cleans better in distilled, soft water. Because detergent does not form scum, it is able to clean better than soap in hard water.

55. How Strong Is Your Cleaner?

When some solid cleaning compounds are put in water, they form bases and can react with an acid to make a neutral solution. The stronger the base, the more acid with which it reacts and the more it reacts in the cleaning process as well. When phenolphthalein is added to a base, it turns pink. When an acid is added to a base, we know when the solution is neutral because the phenolphthalein turns colorless just at that point. With this analytical technique you will investigate the amount of base in several cleaning compounds to compare their relative strengths.

Materials

1. Vinegar (acetic acid, CH_3COOH).
2. Sodium bicarbonate (baking soda, $NaHCO_3$).
3. Sodium carbonate (washing soda or sal soda, Na_2CO_3).
4. Trisodium phosphate (TSP, Na_3PO_4).
5. Sodium borate {borax, $Na_2[B_4O_5(OH)_4]\cdot 8H_2O$}.
6. Phenolphthalein.
7. Buret.
8. Two Erlenmeyer flasks, 125 or 250 mL.
9. Two graduated cylinders, 25 and 100 mL.
10. Balance.

Procedure

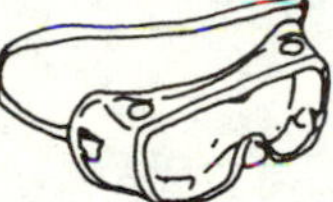

1. Wear goggles. To prepare a 5% solution of one of the cleaners, add 5.0 g of the solid to 95 mL (95 g) of water to make 100 g of solution.
2. Place exactly 50.0 mL of the solution in each of the two flasks.
3. Add 2 drops of phenolphthalein to each flask.
4. Fill the buret with vinegar to the 0.00 mark.
5. Record the reading on the buret as "volume reading: initial".
6. Add vinegar from the buret to the flask until the phenolphthalein JUST turns colorless. Add the vinegar slowly, one drop at a time, near this point.

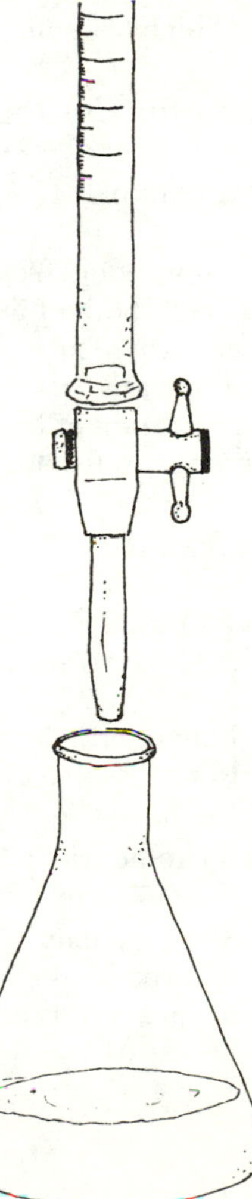

7. Record the reading on the buret as "volume reading: final". Be careful in reading the buret. Why?
8. Determine the volume of vinegar in milliliters used in the titration.
9. Repeat with the other flask of solution. Are your results close to the same? Record all of your data.
10. Repeat the titration with as many cleaners as you can.

Reaction

These solid bases react with water to release hydroxide ion, OH^-, into the solution. The OH^- is the base. The reaction of base with acid, as in titration, is the reaction of OH^- (the base) with H^+ (the acid).

$$OH^- + H^+ \longrightarrow H_2O$$

Bases make good cleaners because they also react with greasy dirt and oil.

Questions

1. Compare the volumes of vinegar that were required to titrate the bases you studied.
2. Which 5% solution contained the most base? Which was next? Compare the class results obtained with all the cleaners.
3. Which 5% solution contained the least base? How do you know?

Notes for the Teacher

BACKGROUND

Most household cleaners are bases because most household dirt is composed of oils and acids, both of which react with the bases. The products can then be rinsed or wiped away. The bases used in this activity are actually salts of weak acids that react in water to tie up the "H" part of the water (HOH) and result in an increased concentration of base (OH^-) in the solution. Titration is an important analytical technique in teaching laboratories and in industrial and research laboratories.

TEACHING TIPS

SAFETY NOTE: Sodium carbonate, trisodium phosphate, and sodium borate must be handled with care. Concentrated solutions can be corrosive and irritating.

1. Phenolphthalein is a good indicator because the salts that result in these titrations dissolve to form a basic solution.
2. Challenge students to figure out how to read the buret. Be sure to tell them that it can be tricky to remember to read from the top down. They will be reading the delivered volume, not the volume of the buret.
3. Students may also wish to learn how to read the buret to the degree of precision allowed by the buret to the nearest 0.02 mL or 0.01 mL. For the comparison purposes of this activity, however, reading the buret so carefully is not really necessary.

4. Another titration that is interesting is the titration of antacid tablets for stomach distress. Use "stomach acid", that is, 0.10 M HCl. Add 100.0 mL of this acid to the weighed tablet. Then titrate with 0.10 M NaOH to find out how much acid was NOT neutralized by the tablet. Because the volume of acid titrated will equal the volume of base in this case, you can subtract the volume of acid neutralized by the sodium hydroxide from the total volume of acid (100 mL) to find the volume of acid neutralized by the tablet.
5. Borax has been an important additive to washing powders for years because it reacts with calcium and magnesium ions, which cause hard water, and removes them from the water so they cannot form soap scum.
6. If your water is acidic, the 5% $NaHCO_3$ solution may not be basic enough to cause phenolphthalein to become pink.

ANSWERS TO THE QUESTIONS

1. Sodium bicarbonate required the least amount of vinegar in this titration.
2. The sodium carbonate required the most acid to become neutral. It was followed by trisodium phosphate and then borax.
3. Sodium bicarbonate contained the least amount of base; we know this because it required the least amount of vinegar to become neutral.

56. Cleaning Compounds and Dirty Surfaces

Cleaning the surfaces of our stoves, floors, windows, and sinks means finding the right chemicals to dissolve the dirt. Some dirt has grease in it, and some dirt has minerals in it. Different kinds of chemicals are needed to clean these two kinds of dirt. In this activity you will find out what kinds of cleaning chemicals work best for different kinds of dirt.

Materials

1. Cleaning chemicals: window cleaner, sink scrubber, ammonia, detergent, soap, vinegar, and other household cleaners. DO NOT USE BLEACH.
2. Powdered calcium carbonate (limestone, $CaCO_3$).
3. Vegetable oil.
4. Wide-range pH paper.
5. Test tubes.
6. Stirring rods.

Procedure

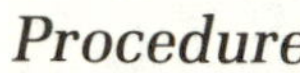

1. Which of the cleaners do you think contain acids or bases? Which are neutral? Test each one with pH paper or litmus paper. Record your results.
2. Select an acidic cleaner and a basic cleaner to investigate.
3. Obtain samples of the following common household substances that represent soiled surfaces: 1) oil, which represents greasy dirt from cooking, and 2) calcium carbonate, which represents mineral deposits.
4. Predict whether each of the soil types is acidic, basic, or neutral. Test with the pH paper. Record your results.
5. For each cleaner to be tested, "soil" two test tubes by coating the inside of one with 1 g of oil and the inside of the other with 1 g of calcium carbonate. One gram of oil is about 1 mL. Estimate the amount, but use the same amount for each test.
6. Using about one-fourth of a test tube of each liquid cleaner or about 1 g of each solid cleaner in one-fourth of a test tube of water, test each of the cleaners with each of the soil types. Record your observations on a data table.
7. Determine which cleaner is able to react with each type of soil.

Reaction

Household dirt is usually either greasy dirt from cooking, street oils, and oily skin, or it is mineral deposits from soap scum, minerals in the water, or dirt from mud and dust. When oil reacts with a base, it forms a type of soap that dissolves in water and is washed away.

When minerals react with an acid, they are dissolved in the solution and then washed away.

Questions

1. Comparing results with the class, list all basic cleaners and all acidic cleaners.
2. Are the household soils acidic or basic?
3. Why do you think the successful cleaners were able to remove the tested soils?
4. Would there be any advantage to the formulation of a cleaner that contained both an acid and a base?

Notes for the Teacher

BACKGROUND

Cleaning around the house can be accomplished with a few well-chosen chemicals in addition to soap and detergent. Of particular use are two bases, ammonia (NH_3) and sodium carbonate (Na_2CO_3), for cleaning greasy dirt and an acid, vinegar, for cleaning soap scum and scale and encrusted minerals. Commercial products are combinations of these and other acids and bases as well as soaps, detergents, fragrances, and colorings. Some of the foamy cleaners even contain both an acid and a carbonate that react to produce foam but result in no acid or base to actually do the cleaning.

TEACHING TIPS

SAFETY NOTE: Be careful with some concentrated cleaners. Even though these cleaners are household chemicals, students should be cautioned to handle them as they would other chemicals. Avoid touching the solids or solutions.

1. Choose any cleaners. Have students bring them from home, including the label. Avoid bleach because reactions could result that produce chlorine gas and other toxic and explosive gases. Also avoid drain cleaner (lye). It is a strong base and is corrosive.
2. This activity provides good practice in constructing a data table. You might require that the students make their data tables before they begin the investigation.
3. Students can suggest other cleaning demands to test the cleaners, for example, baked-on grease and bathtub rings. They can do some controlled investigation at home.
4. Another class of cleaners you may wish to include is that of organic solvents, which includes alcohols and petroleum distillates. Alcohols are present in some household cleaners to dissolve oils, and petroleum distillates remove waxes.
5. Students may want to interview homemakers and those who operate cleaning services to determine which chemicals are in widest use. They may also want to discover the minimum number of chemicals that will serve the maximum number of needs. For example, a mixture of table salt and vinegar will clean copper oxides from copper cookware. (See Activity 58, “Removing Tarnish from Silver”.) Liquid detergent placed in bath water as it is running will eliminate bathtub rings. Why? (See Activity 54, “Preparing a Detergent”.) A bowl of ammonia placed in the oven overnight will dissolve much of the grease, which can be wiped off in the morning.
6. See Activity 55, “How Strong Is Your Cleaner?”
7. Never mix BLEACH AND AMMONIA because poisonous and explosive chloramines are produced:

$$2NH_3(aq) + OCl^-(aq) \longrightarrow H_2O_2(\ell) + \underset{\text{chloramines}}{\text{mixture}(NCl_3, NHCl_2, NH_2Cl)}$$

ANSWERS TO THE QUESTIONS

1.
   - Basic: soaps, detergents, ammonia, trisodium phosphate, sodium carbonate, sodium bicarbonate, calcium carbonate, drain cleaners, oven cleaners, and borax.
   - Acidic: vinegar, phosphoric acid, potassium hydrogen phosphate, scale removers.

2. Calcium carbonate and soap scum are basic. Oil and grease do not react with water and are not either acidic or basic.
3. Acids react with bases and carbonate. Bases react with oils and greases.
4. No, acid reacts with base to make water and a salt.

57. Smelling Salts, Fertilizer, and Ammonia Gas

Where can you find ammonia lurking in a compound? Ammonia gas is easy to identify by its special odor and its ability to turn red litmus paper to blue. You will search for ammonia in some common compounds.

Materials

1. Household ammonia [$NH_3(aq)$].
2. Ammonium chloride solution [$NH_4Cl(aq)$].
3. Ammonium carbonate and/or smelling salts [$(NH_4)_2CO_3$].
4. Calcium oxide solid [$CaO(s)$].
5. Sodium hydroxide solution [$NaOH(aq)$], 1 M.
6. Egg white.
7. Garden fertilizer.
8. Red litmus paper.
9. Test tubes, about five per team.
10. Burner.

Procedure

1. Test household ammonia.
   A. Obtain a test tube containing a small amount of household ammonia.

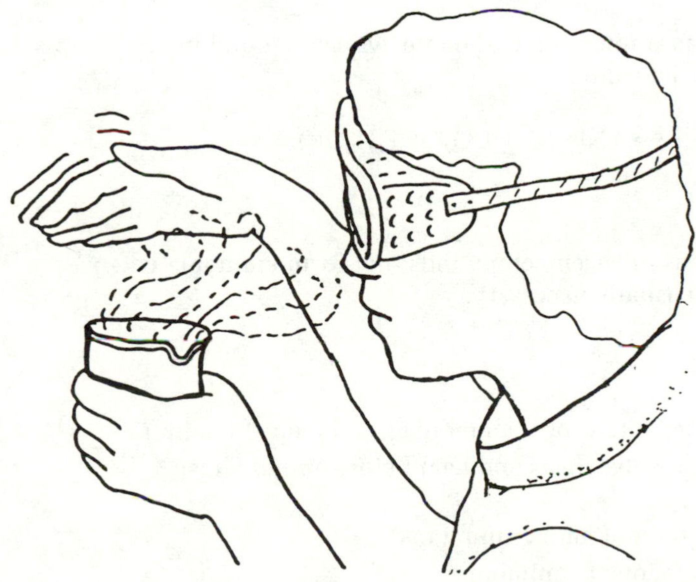

   B. Fan a breeze toward you over the mouth of the test tube to detect the odor. NEVER place your nose or face directly over any odorous chemical.
   C. Hold a moist piece of red litmus paper over the test tube. What happens?
2. Test ammonium chloride.
   A. Place a few milliliters of the ammonium chloride solution in another test tube.
   B. Perform the test in Procedures 1B and 1C; warm the solution if necessary.

C. Try another test: Add about 1 mL of sodium hydroxide solution to the tube containing ammonium chloride. Caution: Sodium hydroxide is a strong base. Perform the test in Procedures 1B and 1C; warm the solution if necessary. Do you detect ammonia molecules?

3. Test other substances.

   A. Test about ⅛ tsp of smelling salts or ammonium carbonate, which is found in smelling salts, by performing the test in Procedures 1B and 1C. Is ammonia present?

   B. Test about ⅛ tsp of garden fertilizer by adding 1 mL of sodium hydroxide solution, then performing the test in Procedures 1B and 1C. Warm if necessary. Is ammonia present?

   C. Test egg white by the following reaction: Place 1 mL of sodium hydroxide solution in a test tube. Add about ⅛ tsp of solid calcium oxide. Mix. Add 1 mL of egg white. Test for ammonia by performing the test in Procedures 1B and 1C. Warm if necessary.

Reactions

Ammonia (NH_3) is a gas that dissolves readily in water. Some of the gas escapes from an ammonia solution; thus, you can smell it, and it can react with litmus. When litmus turns blue, it shows the presence of a base. Ammonia is a base. Ammonium salts react with strong bases such as sodium hydroxide to produce ammonia:

$$NH_4Cl(aq) + NaOH(aq) \longrightarrow \underset{\text{ammonia}}{NH_3(g)} + H_2O(\ell) + NaCl(aq)$$

Some ammonium salts such as ammonium carbonate, which is found in smelling salts, decompose without added base:

$$(NH_4)_2CO_3(s) \longrightarrow \underset{\text{ammonia}}{2NH_3(g)} + CO_2(g) + H_2O(\ell)$$

Protein, like egg white, contains nitrogen compounds that form ammonia when treated with a strong base. Heat is usually necessary.

Questions

1. Which substances did you observe to evolve ammonia gas the most readily?
2. Which substances require other substances and heat before ammonia is formed?
3. Why can you smell ammonia, but not ammonium ions?
4. How do you convert ammonium ions to ammonia?
5. What other compounds do you know of that form or contain ammonia?

Notes for the Teacher

BACKGROUND

Ammonia is a common gas that has a sharp odor. Most people are familiar with it in household cleaning compounds such as oven cleaners and window cleaners. Ammonium salts are found in fertilizers and the old-fashioned smelling salts. This activity is a good way for students to become familiar with ammonia and to discover the difference between ammonia gas (NH_3) and the ammonium ion, NH_4^+.

SOLUTIONS

1. Household ammonia [NH_3(aq)]: Use directly from the bottle.
2. Ammonium chloride solution [NH_4Cl(aq)]: Add 10 g of NH_4Cl to 200 mL of water.
3. Sodium hydroxide solution [NaOH(aq)]: Add 10 g of NaOH to 200 mL of water.

TEACHING TIPS

SAFETY NOTE: Use care with ammonia and sodium hydroxide.

1. This activity has a number of important safety requirements. It is a good one for learning technique. Be sure to demonstrate the proper way to detect odor (see Procedure 1). Students should be very careful with sodium hydroxide solution. Check that every student is wearing safety goggles. Gentle heating in a water bath, if needed, is recommended to avoid bumping the solutions out of the tubes.
2. Ammonia is produced industrially by the Haber process, in which nitrogen from the air and hydrogen (usually from natural gas and water) are combined directly in the presence of a catalyst. Scientists estimate that without this process (developed in 1910), the world population would have leveled off by now because of a lack of ammonia to make fertilizer. Haber ammonia was first used in Germany to make explosives during World War I. Have your students learn more about Fritz Haber, the scientist who discovered this process.
3. Ammonia is also detected in wet baby diapers because the simple urea molecules decompose to simpler ammonia molecules. Fish excrete ammonia directly through their gills.
4. Procedure 2C is the standard qualitative analysis for ammonium ions.
5. The structure of ammonia (NH_3) is the following:

```
      ..
      N
   H/ | \H
      H
```

The structure of the ammonium ion (NH_4^+) is the following:

```
 ┌    H    ┐ +
      |
      N
     /|\
 └ H  H  H ┘
```

ANSWERS TO THE QUESTIONS

1. Ammonia solution and ammonium carbonate evolve ammonia gas most readily.
2. Ammonium chloride and egg white require treatment.
3. Ammonia (NH_3) is a gaseous molecule that forms a solution with water; however, it can escape from the solution into the gaseous phase. Ammonium (NH_4^+) is a positive ion that remains dissolved in water as part of a salt.
4. React NH_4^+(aq) with OH^-(aq) to produce NH_3(aq). NH_3(g) and $H_2O(\ell)$ then are formed.
5. Animals excrete urea, which decomposes to form ammonia; fish excrete ammonia directly into water.

58. Removing Tarnish from Silver

Items made of silver will tarnish if they are exposed to air because the silver reacts with hydrogen sulfide gas in the air. Silver will also react with other sulfides such as those in egg yolks and in rubber. Black silver sulfide is formed. Silver sulfide is the chemical name for "tarnish". You can be a laboratory chemist at home and chemically remove the tarnish using a solution of baking soda and some aluminum foil.

Materials

1. Tarnished silver item (*or* silver item, egg yolk, and rubber band).
2. Baking soda ($NaHCO_3$) or washing soda (Na_2CO_3), 7–8 Tbsp.
3. Scrap aluminum, aluminum foil.
4. Large pan.
5. Tongs.
6. Burner or hot plate.

Procedure

1. Obtain a tarnished silver item or create one as follows:
   A. Wrap a rubber band around one end of a silver item.
   B. Dip the other end into an egg yolk.
   C. Set the item aside until the next day.
2. Wash the tarnished item and examine the tarnished areas.
3. Add 7–8 Tbsp of baking soda or washing soda to 1–2 L of water.
4. Heat the solution in a large pan. Do not boil.
5. Place the scrap aluminum on the bottom of the pan.
6. Place the silver item on the aluminum foil or scraps.
7. Heat the pan almost to boiling.
8. After a few minutes remove the item and rinse it in running water.

Reactions

Silver, Ag, reacts with hydrogen sulfide, H_2S, and oxygen, O_2, in the air to form silver sulfide, Ag_2S, which is the black tarnish:

$$4Ag(s) + 2H_2S(g) + O_2(g) \longrightarrow \underset{\text{black}}{2Ag_2S(s)} + 2H_2O(\ell)$$

Sulfides present in egg yolks and in rubber react with silver in a similar manner.

The chemical cleaning of silver is an electrochemical process in which electrons move from aluminum to silver in the tarnish. This process causes the silver atoms to regain their original form:

$$2Al(s) + \underset{\text{black}}{3Ag_2S(s)} + 6H_2O(\ell) \longrightarrow \underset{\text{silver}}{6Ag(s)} + 2Al^{3+}(aq) + \underset{\text{colorless}}{6OH^-(aq)} + 3H_2S(g)$$

The sodium bicarbonate helps to remove the coating of aluminum oxide to allow the aluminum to react with the silver and provides a conductive ionic solution in which the electrochemical process can occur.

Questions

1. List each observation of a chemical reaction.
2. What happens to the silver atoms when silver tarnishes?
3. What happens to the silver atoms when they react with aluminum?
4. What do you think happens to the aluminum atoms?

Notes for the Teacher

BACKGROUND

Metals react with gases in the air to form corrosion products. Iron reacts in moist air to form rust (iron oxide, Fe_2O_3). Copper reacts with oxygen, carbon dioxide, and other gases to form a protective *patina*, the green coating on copper. Aluminum reacts with oxygen to form a thin, tough film of aluminum oxide, which protects the metal from further reaction. Silver reacts with hydrogen sulfide, a gas produced naturally by the decomposition of organic matter in the soil. The resulting tarnish is black. In this activity you will produce the tarnish and then react aluminum with the silver to remove the silver sulfide tarnish. Aluminum will spontaneously transfer electrons to silver. This reaction results in metallic silver and dissolved aluminum.

TEACHING TIPS

SAFETY NOTE: Be careful with the hot water!

1. Hydrogen sulfide gas is also a product of fuel combustion. Does silver tarnish faster now than it did 200 years ago?
2. Silver is cleaned physically with the mild abrasive in silver polish; however, this abrasive actually removes some of the silver. Silver is cleaned chemically in this activity by re-forming the original metallic silver.
3. Additional baking soda may be added to the solution as you continue the cleaning of several pieces.
4. Use aluminum foil wrap or pie pans, or use an aluminum pan.
5. You should be able to detect the rotten egg odor of hydrogen sulfide gas as it is released during the reaction with aluminum. It is a small amount.
6. Students may be introduced to the ideas of oxidation and reduction and electrode potentials by this reaction.
7. The aesthetic drawback to the reaction is that all of the tarnish may be removed, including that which may be desired to bring out the design in a pattern.
8. This activity can be performed in aluminum pots on the kitchen stove at home.

ANSWERS TO THE QUESTIONS

1. Color change and gas formation.
2. Silver is oxidized; each atom loses one electron.
3. Silver is reduced; each atom gains one electron.
4. Aluminum is oxidized; each atom loses three electrons.

59. Making Artist's Paints from Milk

Most paints consist of colored pigments combined with a binder to keep them soft and to allow them to be spread on paper or canvas. In this activity we will chemically produce several colored pigments. We will then prepare the important binder by precipitating the protein, casein, from skim milk. This process is how some waterproof artist's paints are made commercially. The paints you make in this activity can be used to paint a picture.

Materials

1. Skim milk (can be made from milk powder).
2. Vinegar (acetic acid, CH_3COOH).
3. Calcium chloride, $CaCl_2$.
4. Sodium carbonate (washing soda), Na_2CO_3.
5. Ammonium iron(III) sulfate (also called ferric ammonium sulfate), $NH_4Fe(SO_4)_2 \cdot 3H_2O$.
6. Sodium silicate (water glass) solution, $Na_2SiO_3(aq)$.
7. Potassium hexacyanoferrate(II) (also called potassium ferrocyanide), $K_4Fe(CN)_6$.
8. Cobalt chloride, $CoCl_2$.
9. Powdered charcoal, C.
10. Small test tubes with stoppers to fit, one per pigment.
11. Filter paper.
12. 250-mL beaker.
13. Stirring sticks.
14. Knife for chopping.
15. Mortar and pestle or kitchen blender.
16. Evaporating dish.
17. Wooden splint or spatula.

Procedure

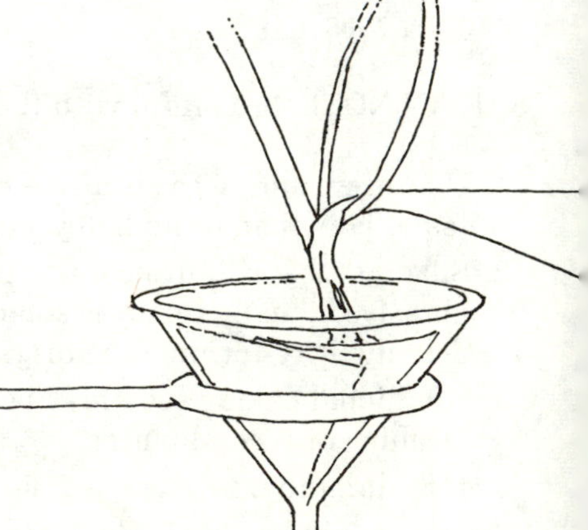

1. Make the pigments.
   A. WHITE pigment:
      1. Place 0.3 g of calcium chloride in a test tube. Fill the tube half full with warm water. Shake the tube gently until the calcium chloride dissolves.
      2. Add 0.3 g of sodium carbonate to the tube. Stopper the tube and shake it thoroughly.
      3. Filter the solution. Discard the liquid filtrate and save the precipitated pigment on the filter paper.

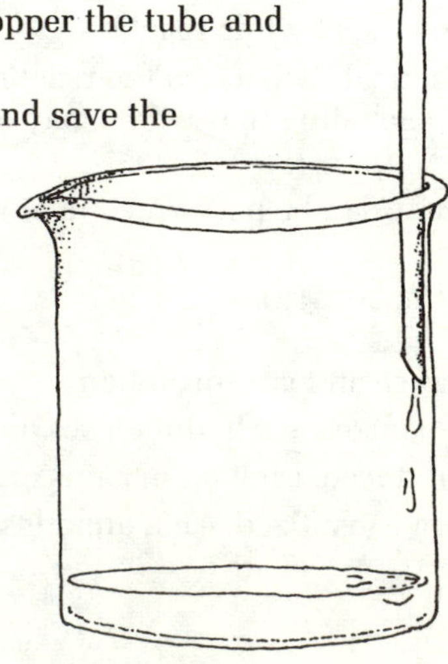

B. GREEN pigment:

1. Place 0.3 g of potassium hexacyanoferrate(II) in a test tube. Fill the tube half full with warm water. Shake the tube until the solid dissolves.

2. Add 0.2 g of cobalt chloride to the tube. Stopper the tube and shake it thoroughly. Avoid touching the cobalt chloride. It is toxic.

3. Filter the solution. Discard the liquid filtrate and save the precipitated pigment on the filter paper.

C. BROWN pigment:

Follow the procedure described for the WHITE pigment, but instead dissolve 0.2 g of ammonium iron(III) sulfate in a tube half-filled with warm water. Then add 0.2 g of sodium carbonate. Shake and filter.

D. BLUE pigment:

Follow the procedure described for the WHITE pigment, but instead dissolve 0.2 g of ammonium iron(III) sulfate in a tube half-filled with warm water. Then add 0.2 g of potassium hexacyanoferrate(II). Shake and filter.

E. ORANGE pigment:

Follow the procedure described for the WHITE pigment, but instead dissolve 0.2 g of ammonium iron(III) sulfate in a tube half-filled with warm water. Then add 1.0 mL of sodium silicate solution. Shake and filter.

F. ROYAL BLUE pigment:

Follow the procedure described for the WHITE pigment, but instead dissolve 0.2 g of cobalt chloride in a tube half-filled with warm water. Then add 1 mL of sodium silicate solution. Shake and filter.

G. LAVENDER pigment:

Follow the procedure described for the WHITE pigment, but instead dissolve 0.2 g of cobalt chloride to a tube half-filled with warm water. Then add 0.2 g of sodium carbonate. Shake and filter.

2. Make the casein binder.

A. Fill a 250-mL beaker three-fourths full with skim milk (or make the skim milk from powdered milk).

B. Gently heat the milk until it begins to boil.

C. Turn off the burner and carefully remove the beaker of hot milk.

D. Slowly, while stirring, add 10 mL of vinegar. Notice the formation of a precipitate.

E. Let the milk sit for a few seconds. If the liquid is still cloudy, add a little more vinegar while stirring. Continue to add vinegar in this manner until the liquid is almost clear.

F. Carefully decant the liquid from the beaker and remove water from the thick precipitate by wrapping it in a layer of cheesecloth or filter paper and gently squeezing it.

G. This precipitate is CASEIN. Chop the casein into small pieces and dry it. Then grind it into a fine powder with a mortar and pestle or a kitchen blender.

3. Make the paints.

A. Place a small amount of casein (enough to cover a penny) in an evaporating dish. Add just enough water to make a thick paste.

B. Add about the same amount of the white pigment that you prepared in Procedure 1A.

C. Using a wooden splint or spatula, mix the casein and the white pigment until the desired white color is obtained. You may add more casein or pigment.
D. Repeat the process, adding the prepared pigments to about the same amount of casein.
E. Prepare BLACK paint by adding some powdered charcoal to a small amount of casein.
F. Paint a picture!

Reactions

The following represent the most probable reactions to produce the colored pigments:

A. WHITE

The white pigment is calcium carbonate (a form of chalk):

$$\underset{\text{colorless}}{Ca^{2+}(aq)} + \underset{\text{colorless}}{CO_3^{\ 2-}(aq)} \longrightarrow \underset{\text{white}}{CaCO_3(s)}$$

B. GREEN

The green pigment is cobalt hexacyanoferrate(II):

$$\underset{\text{pink}}{2Co^{2+}(aq)} + \underset{\text{pale yellow}}{[Fe(CN)_6]^{4-}} \longrightarrow \underset{\text{gray-green}}{Co_2Fe(CN)_6}$$

C. BROWN

The brown pigment is iron(III) hydroxide. The hydroxide forms when the carbonate ion reacts with water:

$$CO_3^{\ 2-}(aq) + H_2O \longrightarrow HCO_3^{\ -}(aq) + OH^-(aq)$$

$$\underset{\text{pale yellow-green}}{Fe^{3+}(aq) + 3OH^-(aq)} \longrightarrow \underset{\text{brown}}{Fe(OH)_3(s)}$$

D. BLUE

The blue pigment is called Prussian blue:

$$\underset{\text{pale yellow-green}}{K^+ + Fe^{3+}(aq)} + \underset{\text{pale yellow}}{[Fe(CN)_6]^{4-}(aq)} \longrightarrow \underset{\text{dark blue}}{KFe_2(CN)_6(s)}$$

E. ORANGE

The orange (possibly buff or red) is iron(III) hydroxide mixed with iron(III) silicate. There is an excess of hydroxide ions in a sodium silicate solution.

$$\underset{\text{pale yellow}}{Fe^{3+}(aq)} + \underset{\text{colorless}}{3OH^-(aq)} \longrightarrow \underset{\text{pale orange}}{Fe(OH)_3}$$

F. BLUE

The blue pigment is cobalt silicate:

$$\underset{\text{pink}}{Co^{2+}(aq)} + \underset{\text{colorless}}{SiO_3^{\ 2-}(aq)} \longrightarrow \underset{\text{royal blue}}{CoSiO_3(s)}$$

G. LAVENDER

This pigment is cobalt carbonate:

$$\underset{\text{pink}}{Co^{2+}(aq)} + \underset{\text{colorless}}{CO_3^{2-}(aq)} \longrightarrow \underset{\text{pale purple}}{CoCO_3(s)}$$

Questions

1. How is the protein casein chemically separated from milk?
2. Identify the binder and the pigment in each paint that you produced.
3. What properties should the pigments have in order to be useful as paints?

Notes for the Teacher

BACKGROUND

This activity contains a lot of chemistry. Students will be able to carry out several chemical reactions to produce pigments that can be added to casein, which is separated from milk. Casein is the primary protein in milk. Casein and other proteins are denatured and chemically changed by the action of acids, bases, alcohols, heavy metal ions (e.g., mercury and lead), detergents, and many other chemicals. Casein, when separated from milk, forms a binder for the solid-colored pigments also produced in this activity. Students should compare their products to commercially available artist's paints and try them.

SOLUTIONS

Use commercial Na_2SiO_3 (water glass) or prepare a saturated solution of the solid.

TEACHING TIPS

1. You may wish to collect all of the casein from the class, place it in a blender with a little water, and make enough of the paste for the class.
2. Commercial treatment of 100 lb of milk will yield 3 lb of casein.
3. When dissolved in an alkaline solution, this same casein will become sticky and can be used as glue. See Activity 50 to make glue from casein.
4. In addition to its use as a paint binder and glue, casein is also used as a binder of clay in paper coating of high-quality paper, as a top coat in fine leather for waterproofing and protection, and as an additive to wine for clarification.
5. Casein is also used to make some plastics. Buttons are often made of casein plastics.
6. Acid precipitates casein from solution in milk. This process occurs naturally when the lactic acid in milk causes the precipitation of casein when the milk has become sour.
7. Casein is the primary source of protein for milk-fed infants. Hydrochloric acid in the stomachs of infants changes milk to casein for more efficient digestion.
8. Encourage students to try other pigments and chemicals with the casein binder. For example, powdered blackboard chalk added to casein will produce a nice white paint. Iron(III) oxide produces a red pigment when mixed with casein.
9. Casein will dry out. Keep it moist by storing it in a plastic bag.
10. The cyanide in ferrocyanide is chemically bound and is not a toxic cyanide.
11. Ammonium iron(III) sulfate is called an *alum*. Alums have the general formula $M^+M^{3+}SO_4{\cdot}12H_2O$, where M^+ is Li^+, Na^+, K^+, or NH_4^+, and M^{3+} is Cr^{3+}, Fe^{3+}, or Al^{3+}.

12. The red color in the mineral garnet is due primarily to $Fe_3{}^{2+}Fe_2{}^{3+}(SiO_4)_3$.
13. Students might find it helpful to have the directions for this activity in chart form.

Pigment	*Reactants*	*Product*
White	0.3 g of $CaCl_2$ 0.3 g of Na_2CO_3	$CaCO_3$
Green	0.3 g of $K_4Fe(CN)_6$ 0.2 g of $CoCl_2$	$Co_2Fe(CN)_6$
Brown	0.2 g of $NH_4Fe(SO_4)_2 \cdot 3H_2O$ 0.2 g of Na_2CO_3	$Fe(OH)_3$
Blue (dark)	0.2 g of $NH_4Fe(SO_4)_2$ 0.2 g of $K_4Fe(CN)_6$	$KFe_2(CN)_6$
Orange	0.2 g of $NH_4Fe(SO_4)_2$ 1.0 mL of Na_2SiO_3	$Fe_2(SiO_3)_3 +$ $Fe(OH)_3$
Blue (royal)	0.2 g of $CoCl_2$ 1.0 mL of Na_2SiO_3	$CoSiO_3$
Lavender	0.2 g of $CoCl_2$ 0.2 g of Na_2CO_3	$CoCO_3$

ANSWERS TO THE QUESTIONS

1. The casein was precipitated by the action of acid.
2. The binder is casein. The pigments are given in the Reactions section.
3. The pigment should have color. It should also be a solid and should be stable. The pigments are not really dissolved in the binder; instead, they are suspended in it.

60. Preparing Sodium Hydrogen Carbonate (Baking Soda)

Sodium hydrogen carbonate, also called sodium bicarbonate, bicarbonate of soda, and baking soda, is an important chemical. Hundreds of thousands of tons are produced each year for use in baking and in producing other chemicals. In this activity, you will prepare sodium hydrogen carbonate and test it by comparing its reactions with those of baking soda sold commercially.

Materials

1. Sodium chloride, NaCl.
2. Ammonium carbonate, $(NH_4)_2CO_3$.
3. Solid carbon dioxide (dry ice), CO_2.
4. Vinegar (dilute acetic acid), CH_3COOH.
5. pH paper, red litmus paper, or indicator solution.
6. Copper sulfate solution, $CuSO_4$, 1 M.
7. Two beakers, 400 mL.
8. Funnel.
9. Filter paper.
10. Stirring rod.
11. Six test tubes.

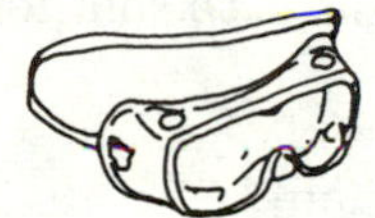

Procedure

1. Preparing sodium hydrogen carbonate
   A. Place 15 g of sodium chloride in a beaker and add 125 mL of water. Stir until the salt dissolves.
   B. While you stir, add 30 g of ammonium carbonate to the solution. Stir until no more solid will dissolve.
   C. After the undissolved ammonium carbonate settles to the bottom of the beaker, carefully pour off the liquid into another large beaker.
   D. Add about 70 g of dry ice, in large chunks, to the solution. Add the pieces of dry ice one at a time. Be sure that they are large enough to fall to the bottom of the solution. Stir the solution frequently.
   E. When all of the dry ice has reacted, notice that a white precipitate has formed. This precipitate is sodium hydrogen carbonate.
   F. Set up a filtering apparatus as shown in the diagram. Place the original beaker beneath the funnel and carefully filter the solution, transferring all of the precipitate to the funnel.
   G. Carefully remove the filter paper and place it in an area where the precipitate can dry in the air.

2. Testing the sodium hydrogen carbonate
   A. Label six large test tubes: three of them "Control" and three of them "Experiment". Place them in pairs, one from each group, in a test tube rack or in a large beaker.
   B. Place a pea-sized amount of sodium bicarbonate (preferably from a fresh box) into each of the control tubes.
   C. Place an equal amount of the product you prepared into each of the experiment tubes.
   D. Add 5 mL of water to each tube and gently shake the tube until all of the solid is in solution.

E. Test the pH of the first set of tubes with test paper or solution. How do they compare? Record your results.
F. Add 5 mL of vinegar to the second set of tubes. What do you observe? How does the reaction in the two tubes compare? Record your observations.
G. Add 5 mL of copper sulfate solution to each tube in the third pair. What do you observe? Record your observations.

Reactions

1. The ammonium carbonate reacts in water to form ammonium bicarbonate and ammonia gas:

$$(NH_4)_2CO_3\ (s) \longrightarrow NH_4HCO_3\ (aq) + NH_3\ (g)$$

The ammonium hydrogen carbonate ionizes to produce the hydrogen carbonate (or bicarbonate) ion, HCO_3^- and the ammonium ion, NH_4^+. The dry ice forces carbon dioxide into the solution with sodium chloride and ammonium bicarbonate and also cools the solution. Sodium hydrogen carbonate that forms is only slightly soluble in cold water, so it precipitates.

$$Na^+ + HCO_3^-\ (aq) \longrightarrow NaHCO_3\ (s)$$

2.

A. The bicarbonate ion, HCO_3^-, is a stronger base than it is an acid, so solutions of salts of bicarbonate are weakly alkaline.

$$HCO_3^-\ (aq) + H_2O \longrightarrow H_2CO_3\ (aq) + OH^-\ (aq)$$

The pH of a sodium bicarbonate solution is about 8.4.

B. The acetic acid in vinegar reacts with the bicarbonate ion to release carbon dioxide gas.

$$H_3O^+\ (aq) + HCO_3^-\ (aq) \longrightarrow CO_2\ (g) + 2H_2O\ (\ell)$$

C. Copper(II) sulfate forms an acid solution with water. This causes the release of carbon dioxide gas. The copper(II) ion forms a blue, insoluble compound with carbonates.

$$Cu^{2+}\ (aq) + CO_3^{2-}\ (aq) \longrightarrow Cu(II)CO_3\ (s)$$

Questions

1. Why is sodium bicarbonate slightly basic?
2. Why is it necessary to prepare sodium bicarbonate at a low temperature?
3. How does the bicarbonate you prepared compare with the sodium bicarbonate sold commercially?

Notes for the Teacher

BACKGROUND

This activity gives students an opportunity to synthesize a common chemical compound and test it by comparing it to the identical commercial product. Sodium hydrogen carbonate, better known as sodium bicarbonate or baking soda, is an important industrial compound; it is the starting material for the synthesis of many products, including baking powder. See Activity 77, "What's in Baking Powder?"

SOLUTIONS

1. Vinegar is 5% acetic acid. Any dilute acid will do. Caution students to use acids with care and to clean up any spills.
2. Copper(II) sulfate solution is 0.1 M. Dissolve 25 g copper(II) sulfate pentahydrate ($CuSO_4 \cdot 5H_2O$) per liter of solution.

TEACHING TIPS

SAFETY NOTE: Be careful with dry ice. Handle dry ice with heavy gloves and tongs.

1. Students can obtain the product faster by suction filtration.
2. Keep the ammonium carbonate covered. When exposed to air, ammonium carbonate gives off ammonia gas. This product is often used as the smelling salts that arouse people who faint.
3. When an equivalent amount of sodium bicarbonate and sodium hydroxide are mixed, crystals of sodium carbonate decahydrate form upon evaporation of the solution. This hydrate is called "washing soda". It is commonly used as a water softener.
4. If sodium carbonate decahydrate is gently heated, the water is driven off and anhydrous sodium carbonate is formed. This compound (sometimes called "soda ash"), unlike sodium bicarbonate, is strongly basic due to the extensive hydrolysis of the carbonate ion:

$$CO_3^{2-}\ (aq) + H_2O \longrightarrow HCO_3^-\ (aq) + OH^-\ (aq)$$

5. Sodium hydrogen carbonate is mixed with a weak acid to form baking powder. Sodium dihydrogen phosphate is commonly used as a source of weak acid. When these two substances are mixed with water and warmed, carbon dioxide forms and acts as a leavening agent to cause dough to rise:

$$H_3O^+\ (aq) + HCO_3^-\ (aq) \longrightarrow CO_2\ (g) + 2H_2O$$

6. Sodium hydrogen carbonate is made commercially from sodium carbonate or sodium hydroxide. Sodium hydroxide is not suggested for this activity because of its corrosive properties.

$$Na_2CO_3\ (aq) + H_2O\ (\ell) + CO_2\ (g) \longrightarrow 2NaHCO_3\ (s)$$

7. Sodium hydrogen carbonate was made in ancient times by leaching ashes. Early in the 18th century, Nicolas Leblanc prepared it by reacting sodium chloride and sulfuric acid to form sodium sulfate and then heating this with coke and limestone to form sodium carbonate. In 1860, Ernest Solvay, a Belgian chemist, improved the process, and today most sodium carbonate and sodium bicarbonate is produced using the Solvay process. This activity is basically a modification of that process.

ANSWERS TO THE QUESTIONS

1. The bicarbonate ion is a stronger base than it is an acid. It reacts with water molecules to form the hydroxide ion in solutions.
2. Sodium bicarbonate is only slightly soluble in cold solution. In a cold solution, a large amount of sodium bicarbonate will not dissolve and can be recovered as a solid.
3. They are both basic; they have about the same pH (about 8.5). They both react with vinegar to produce carbon dioxide gas; they both react with copper(II) sulfate to produce carbon dioxide gas and copper carbonate, a light blue precipitate.

61. Decomposing Baking Soda

Baking soda, which is also known as sodium bicarbonate ($NaHCO_3$), is a substance commonly used in baking to cause biscuits to rise. The biscuits rise because the sodium bicarbonate liberates carbon dioxide gas when it reacts with an acid. What happens when pure sodium bicarbonate is heated? A gas is also produced. You will test for this gas and try to determine the other products that remain.

Materials

1. Baking soda (sodium bicarbonate, $NaHCO_3$).
2. Limewater [saturated calcium hydroxide ($Ca(OH)_2$) solution].
3. Hard glass test tube (e.g., Pyrex).
4. Burner.
5. Litmus paper.
6. Balance.

Procedure

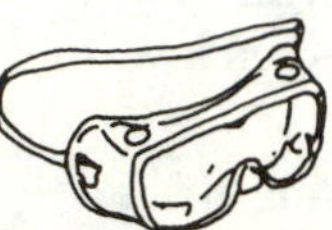

1. Find the mass of a test tube. Record the mass.
2. Add a little sodium bicarbonate to the test tube.
3. Find the mass of the tube plus the sodium bicarbonate.
4. Heat the tube strongly for 3 min.
5. When the tube is cool, find the mass again.
6. Compare the mass of the solid product left in the test tube with the mass of the original sodium bicarbonate.
7. Look at these three mass ratios and find the one closest to your result. This mass ratio provides a clue to the identity of the product.

   Mass ratios of possible products compared to sodium bicarbonate:

$$\frac{0.63 \text{ g of product}}{1.0 \text{ g of sodium bicarbonate}} = Na_2CO_3$$

$$\frac{0.24 \text{ g of product}}{1.0 \text{ g of sodium bicarbonate}} = NaOH$$

$$\frac{0.37 \text{ g of product}}{1.0 \text{ g of sodium bicarbonate}} = Na_2O$$

8. Methods to find out what other substances were produced:
   A. Place a fresh sample of sodium bicarbonate in a test tube, attach a stopper with tubing leading to a beaker of limewater, and heat the tube again. If carbon dioxide is produced, the carbon dioxide will react with the limewater to turn it milky.
   B. Place a fresh sample of sodium bicarbonate in a test tube, attach the rubber stopper but remove the rubber tubing, and heat the tube again. Hold a piece of cool glass above the glass tube from the rubber stopper. Look for water to condense on the cool glass.

Reaction

When sodium bicarbonate is heated, it decomposes. You have been a detective in searching for the three products of the decomposition reaction:

$$NaHCO_3 \xrightarrow{\text{heat}} \underline{\qquad} + \underline{\qquad} + \underline{\qquad}$$

Questions

1. What evidence do you have that the sodium bicarbonate was decomposing when it was heated?
2. What was the mass of the sodium bicarbonate in the test tube?
3. What was the mass of the solid product left in the test tube?
4. What is the ratio of the mass of the product to the mass of the sodium bicarbonate? (Divide the mass of the product by the mass of the sodium bicarbonate).
5. How close is your ratio to one of the ratios given in the Procedures section?
6. What other products were you able to identify?

Notes for the Teacher

BACKGROUND

This reaction is a typical decomposition reaction. It is useful for introducing students to types of reactions, percent composition, writing equations, and reactions of carbon dioxide.

SOLUTION

Prepare the calcium hydroxide solution (limewater) by stirring up to ½ cup of calcium oxide or calcium hydroxide into 1 L of water. Allow the solution to stand so that all of the undissolved calcium hydroxide can settle. Pour off the clear solution. It is a good idea to prepare the limewater the day before you do the activity.

TEACHING TIPS

1. This activity is a good way to practice data keeping on a data table. Allow students to figure out how to find the mass of the substance in the tube (by subtraction).
2. Some students may need to be led to the calculation to find their comparison ratio. Divide the mass of the product by the mass of the sodium bicarbonate to get the ratio:

$$\frac{\text{mass of product}}{\text{mass of sodium bicarbonate}}$$

3. The ratio should be approximately 0.6. The product is sodium carbonate (Na_2CO_3). We see that 2 mol of sodium bicarbonate yields 1 mol of sodium carbonate, 1 mol of water, and 1 mol of carbon dioxide.
4. Sodium carbonate is washing soda.
5. Carbon dioxide is detected when it reacts with calcium hydroxide to form the white precipitate, calcium carbonate ($CaCO_3$).

6. To save time, you may want to demonstrate Procedures 8A and 8B.
7. The balanced equation is the following:

$$2NaHCO_3(s) \longrightarrow Na_2CO_3(s) + H_2O(g) + CO_2(g)$$

Water will condense on a cold glass plate. You may need to use fresh sodium bicarbonate for Procedure 8B.

ANSWERS TO THE QUESTIONS

1. Two gases were formed: One reacted with limewater to produce a precipitate, and the other condensed to form a colorless liquid.
2. Result of subtraction of mass of tube from mass of tube plus solid.
3. See Answer 2.
4. See Teaching Tip 2.
5. Ratios are generally close to 0.6.
6. Carbon dioxide and water.

Chemistry and the Environment

62. Purification of Water

Most of the water that we use in our homes comes from rivers and lakes. This water contains soil, toxic substances, harmful bacteria, and other impurities that must be removed before we can drink it. When water is treated for drinking, it is first allowed to settle in large tanks to allow large particles to fall to the bottom. The water is then treated to remove other suspended matter to produce clear water. In this activity we will chemically process muddy water to produce clear water. We would need additional treatment of the clear water before it could become drinking water.

Materials

1. Two large beakers or jars.
2. Soil.
3. Lime (calcium oxide, CaO) solution.
4. Alum {potassium aluminum sulfate [$KAl(SO_4)_2$] or ammonium aluminum sulfate [$NH_4Al(SO_4)_2$]}.
5. Red litmus paper.

Procedure

1. Fill one of the beakers three-fourths full with water. Add 2-3 tsp of soil.
2. Shake or stir until a muddy solution results.
3. Pour half of this muddy water into a second beaker.
4. One beaker will be our control. We will not treat the water in this beaker; we will compare it with the treated beaker.
5. Add 1 dropper full of *lime* solution (or a match-head size amount of solid lime) to the second beaker of muddy water. Stir the mixture thoroughly.
6. Dip a strip of red litmus paper into the solution. If the litmus paper does not turn blue, add more lime solution, a little at a time, until it does.
7. Add 1 dropper full of *alum* solution to the mixture in this beaker. What do you observe?
8. Continue adding alum until a thick, white, gelatinous precipitate forms. Stir the water thoroughly.
9. Place the two beakers side-by-side and observe them for several minutes.
10. Record and discuss your observations.

Reactions

Two reactions produce the white gelatinous precipitate. The first is the reaction of lime (calcium oxide) with water to produce a basic solution:

$$CaO(s) + H_2O(\ell) \longrightarrow Ca^{2+}(aq) + 2OH^-(aq)$$

Red litmus paper turns blue in this basic solution.

The second reaction produces a precipitate when aluminum ions react with the hydroxide:

$$Al^{3+}(aq) + 3OH^-(aq) \longrightarrow Al(OH_3)(s)$$

As the heavy gelatinous precipitate of aluminum hydroxide settles to the bottom of the beaker, it carries the soil and suspended particles with it and leaves clear water.

Questions

1. How does the treated water sample compare with the control (untreated) sample?
2. How do you know that a basic solution was produced when lime was added to the muddy water?
3. Why was it necessary to make the sample basic?
4. What happened when alum was added to the water?

Notes for the Teacher

BACKGROUND

In this activity students can produce a relatively clear water sample from muddy water by a purification process called *settling.* In the usual treatment of river or lake water for human consumption, the larger particles are allowed to fall to the bottom in large tanks. This process is called *sedimentation.* The water is then treated by settling the impurities with a gelatinous precipitate, aluminum hydroxide. This process removes most of the soil and bacteria. Chlorine is added to kill the remaining bacteria. The water is often sprayed in the air (aeration) to speed up the oxidation of dissolved organic substances, and then it is filtered through charcoal to remove the last traces of impurities as well as any odors. Activity 94 shows the use of charcoal to remove impurities.

SOLUTIONS

1. Lime: You can either use solid calcium oxide (CaO) or make a solution by adding 2–3 tsp to 1 L of water. The amount is not really critical. The calcium hydroxide, $Ca(OH)_2$, that forms does not dissolve well, and the excess will settle to the bottom.
2. Alum: Make a solution by adding 2–3 tsp of aluminum sulfate [$Al_2(SO_4)_3$], potassium aluminum sulfate, or ammonium aluminum sulfate to 1 L of water. Again, the amounts are not critical.

TEACHING TIPS

1. If you do not have lime, you can add a small amount of ammonium hydroxide to the water to make it basic.
2. Even after water is finally purified, it still contains small amounts of sodium, potassium, magnesium, fluoride, chloride, and other ions. These ions are useful as minerals for the body.
3. Some ions remaining in water such as iron, calcium, magnesium, chloride, hydrogen carbonate, and sulfate reduce the cleansing power of soap. Water with a high concentration of such ions is called *hard water.* Activity 54 calls for examining some of these ions in hard water and the reactions that they undergo.
4. Let students bring soil samples from home in their own jars.
5. Emphasize the need for a control in this activity.
6. Although the amount of solution added to the water sample is not critical in this activity, a dropper full of lime solution (or a small amount of the solid) and a dropper full of the aluminum sulfate solution should be adequate.
7. You can buy lime and alum at the drugstore and hardware store.

ANSWERS TO THE QUESTIONS

1. The treated sample became clear as the precipitate and dissolved solids settled to the bottom of the beaker. The control sample remained muddy.
2. It was tested with litmus paper. A basic solution causes red litmus to change to blue.
3. A basic solution contains "hydroxide". The hydroxide reacts with the alum to form the heavy gelatinous precipitate called aluminum hydroxide.
4. A thick, white, gelatinous precipitate formed.

63. Corrosion of Iron: The Statue of Liberty Reaction

In 1984, repair work began on the 100-year-old Statue of Liberty because the internal framework of iron "ribs" was corroded. The internal framework was built to help support the copper statue. Cracks developed in the statue and allowed the entry of wet, salty air. When the protective dividers between the iron and copper wore away, the two metals came into contact with each other and were surrounded by wet salty air, which came through the cracks in the statue.

We will experiment with the reactions that occur when iron, copper, water, and salt come together. Does contact with copper cause iron to rust more rapidly? **This activity will take 2–3 days.**

Materials per Team

1. Six pieces of copper such as pure copper pennies (1981 or OLDER) or bare copper wire or sheet.
2. Six pieces of iron such as iron nails, iron wire, or paper-covered iron twist ties.
3. Salts and acids such as table salt and vinegar.
4. Nine small plastic cups.
5. Paper towels.
6. Fine sandpaper or steel wool.
7. Plastic wrap.
8. Rubber bands.

Procedure

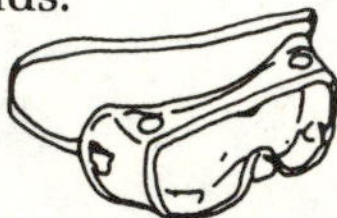

1. Clean the six pieces of copper with the sandpaper or steel wool so that most of the copper oxide tarnish is removed.
2. Clean the six nails with the sandpaper or steel wool also.
3. Using the paper towels, salts, vinegar, and nine plastic cups, create three each of three types of environments to investigate the causes of the corrosion of the iron ribs inside the copper Statue of Liberty:
   A. Three moist, salty, acidic environments.
   B. Three moist environments (use distilled water, if available).
   C. Three dry environments.
4. Inside each of the types of environments, lay the following metals:
   A. Copper piece alone.
   B. Copper piece with a nail on top of it.
   C. Nail alone.
5. Cover each cup tightly with plastic wrap.
6. Observe the nine cups each day to compare the results of the three environments on the metals.
7. Construct a data table with nine spaces for observations.

Reactions

1. In the corrosion reaction, the iron will lose electrons 1000 times faster when in contact with copper. The iron atoms lose electrons to oxygen, which is dissolved in the water:

$$Fe \longrightarrow Fe^{2+}(aq) + 2e^-$$

$$H_2O(\ell) + O_2 + 2e^- \longrightarrow 2OH^-(aq)$$

2. Next, the iron ions combine with the hydroxide ions (OH^-) to form iron(II) hydroxide:

$$Fe^{2+}(aq) + 2OH^-(aq) \longrightarrow Fe(OH)_2(s)$$

3. This iron(II) hydroxide quickly reacts with oxygen in the air to form "rust":

$$2Fe(OH)_2(s) + \tfrac{1}{2}O_2(g) \longrightarrow \underset{\text{rust}}{Fe_2O_3 \cdot H_2O} + H_2O(\ell)$$

Questions

1. Which metal sample corrodes the fastest? Why?
2. What conditions seem to be necessary for the rapid corrosion of the metal?
3. Plan an experiment to test each variable separately to find which is/are most important in promoting the reactions.

Notes for the Teacher

BACKGROUND

The structural integrity of the Statue of Liberty has been threatened because of the corrosion of the iron internal structure. The internal structure was designed by A. G. Eiffel, who designed the famous tower in Paris. The French people presented the citizens of the United States with the Statue of Liberty in 1886 to commemorate friendship between the nations.

The corrosion reaction is speeded up because of the greater tendency of iron to lose electrons than copper. The presence of the copper allows the electrons leaving the iron to flow more rapidly through the copper to the dissolved oxygen. All of the metallic elements differ in their tendency to lose electrons, and when any two are in contact, a flow of electrons occurs. The precious metals (i.e., silver, gold, and platinum) have the least tendency to lose electrons.

TEACHING TIPS

1. Copper pennies work best for this investigation because they are uniform in size and shape. Be sure to use the older ones because new pennies are mostly zinc with a copper coating. The zinc, being another metal, will interfere with the reaction. Zinc has a greater tendency than iron to lose electrons and will corrode, whereas the iron only acts as an electron flow medium. An interesting additional experiment would be to use these new pennies with some of the zinc exposed and in contact with the iron nail and the copper. Can added zinc protect the corrosion of iron? Yes.
2. The paper towels are used to maintain a moist environment in the plastic cup.
3. Small amounts of the salts and vinegar are best. Vinegar is an acid. These substances, dissolved in water, are *electrolytes*, that is, solutions that conduct an electric current. Iron(II), Fe^{2+}, is oxidized to iron(III), Fe^{3+}, when it finally reacts with water to form rust.
4. This activity might be set up as a class research project in which the entire class acts as a team. Small groups may prepare parts of the procedure, and the results for the class as a whole can be pooled together.

5. The data table might look like this:

	Copper Alone	Copper + Nail	Nail Alone
Moist, salty, acidic			
Moist			
Dry			

6. Repeat the experiment, trying as many variables as you can: temperature, other acids, other salts, and other concentrations.
7. An interesting and detailed account of the Statue of Liberty corrosion and repair can be found in *Chem Matters*, Vol. 3, No. 2, April 1985. See Appendix 5.
8. $Fe(OH)_3$ is also an acceptable formula for "rust", but it is better represented as the dehydrated oxide, $Fe_2O_3 \cdot xH_2O$. Because the amount of water varies, $Fe_2O_3 \cdot H_2O$ is only an approximate formula for rust.
9. Discard the plastic cups or label them for use with chemicals.

ANSWERS TO THE QUESTIONS

1. The iron with the copper corrodes the fastest because the copper conducts the electrons away to the solution.
2. Moisture, oxygen, salt, and/or acid and contact with copper.
3. Isolate the acid and the salt to determine which has the greatest effect on the corrosion rate. Investigate the effect of the amount of added salt or acid.

64. Making Paper

Paper production is one of the most important industries in this country. Each year, Americans consume more than 800 lb of paper per person! The world average consumption is only about 80 lb per person. The United States consumes more than one-third of all the paper produced in the world! Paper is made from wood, more specifically from *cellulose*, the polysaccharide found in wood. This polysaccharide is long chains of glucose (simple sugar) molecules. Cellulose molecules form long fibers that are held together in the plant material by *lignin*, a large and complex polymer. In this activity you will digest wood by removing the lignin and recover the cellulose in the form of paper. **This activity will take 2 days.**

Materials

1. Sawdust or a rasp and pieces of lumber (white pine works well).
2. Sodium hydroxide (NaOH) solution, 6 M.
3. Sodium sulfide (Na_2S) solution, 3 M.
4. Laundry bleach.
5. Beakers, 100 and 300 mL.
6. Graduated cylinder, 100 mL.
7. Bunsen burner.
8. Rustproof wire screen.
9. Glass stirring rod.

Procedure

1. Obtain enough fine wood particles to fill a 100-mL beaker by either of the following methods:
   A. Sift some fine sawdust with a wire screen.
   -or-
   B. Use a carpenter's rasp to rub and scrape off fine particles from a large piece of lumber. Sift these particles through a wire screen. Do not use the particles that will not come through the screen.
2. Pour the beaker full of fine sifted wood particles in a 300-mL beaker.
3. Add 100 mL of sodium hydroxide solution. (Be careful when handling this solution. Wear goggles, gloves, and a face shield.) In the hood, add 25 mL of sodium sulfide solution. Avoid contact with sodium sulfide. It is toxic.

4. Heat this mixture until it boils. Boil gently for 15 min.
5. Remove the beaker from the heat and let the wood pulp settle to the bottom. Notice the appearance of the liquid. It contains resins, gums, lignin, and other soluble substances.
6. Carefully pour off the liquid.
7. Fill the beaker with water, stir to mix, and let the pulp settle again. Then pour off the water layer.
8. Repeat this process two or three times to remove the sodium hydroxide and sodium sulfide.
9. To bleach the pulp to remove some of the color, add 200 mL of laundry bleach. This point might be a good place to stop because the pulp should remain in the bleach overnight.
10. To obtain a really white pulp, soak the pulp in 200 mL of laundry bleach overnight several times. However, we will be content with a less-than-white product and continue. Pour off the bleach and wash the pulp several times as you did before.

11. Place a piece of rustproof wire screen over a large beaker or other container. Carefully pour the pulp onto the wire screen. Spread the pulp out into as thin a layer as possible. Leave it there until it dries.
12. You have made a sheet of paper. It will be more like filter paper or blotter paper than notebook paper. Carefully remove the paper from the wire screen and examine it.

Reactions

The first reaction dissolves lignin out of the wood pulp to free the cellulose pulp fibers. The strong alkaline solution helps to degrade the cellulose, and the sodium sulfide reacts with the lignin to remove it from the wood.

Bleach is added to help remove some of the dark color from the small amount of resins, gums, and other materials still in the pulp.

When the treated and bleached pulp is spread out to dry, water is removed. This water removal causes hydrogen bonds to form between cellulose fibers and gives the paper a solid form with considerable strength.

The hydrogen bonds form between the hydrogen atoms of one cellulose chain and the oxygen atoms of another. Cellulose is made of long chains of glucose molecules.

Questions

1. What is cellulose?
2. What holds cellulose fibers together in wood?
3. Describe the process for making paper.

Notes for the Teacher

BACKGROUND

In this activity students can produce a crude form of paper from wood. Paper can be produced by several methods. This method involves digesting the pulp with sodium hydroxide and sodium sulfide to remove the lignin and then drying the cellulose that remains after digestion. By spreading this material on a wire screen, students will be able to make paper. The paper produced will not be the nice thin white paper that the students usually use. However, it is an excellent example of an important commercial reaction. This activity is most effective when spread over 2–3 days.

SOLUTIONS

1. Pulp: Very fine sawdust from a sawmill or cabinet shop works well. You can produce fine wood particles by rasping a piece of white pine that is held securely in place with a vice or by nailing it to heavy boards.
2. Sodium hydroxide: The concentration of this solution is 6 M. Use gloves, goggles, and a face shield. Dissolve 240 g of NaOH in water to make 1 L of solution.
3. Sodium sulfide: The concentration of this solution is about 3 M. Prepare this solution by dissolving 234 g of sodium sulfide to make 1 L of solution. Avoid contact with sodium sulfide. It is toxic.
4. Laundry bleach: This chemical is used to bleach the brown pulp. Any type will work. The active ingredient in "chlorine" bleach is sodium hypochlorite.

TEACHING TIPS

SAFETY NOTE: Be careful with caustic NaOH and toxic Na_2S. Students should wear gloves and a face shield as well as goggles. They should also work in the hood.

1. Cellulose is a polysaccharide consisting of chains of glucose molecules (a polymer). The structure of cellulose is the following:

CH_2OH CH_2OH CH_2OH
H O H O H O
~O H OH H H O H OH H H O H OH H H O~
H OH H OH H OH

2. There are more than 12,000 types of paper!
3. The conversion of wood to pulp produces large amounts of waste solutions. Foul-smelling hydrogen sulfide, methanethiol (methyl mercaptan), and dimethyl sulfide account for most of the pollution. The bleaching process requires about 20,000 gal of water for every ton of pulp processed. The discharge of this colored waste water is a major source of water pollution.
4. We are using the alkali method of paper production in this activity. The sulfite method is more commonly used in industry. In this method, bisulfite (from sodium sulfite or calcium bisulfite) reacts with lignins to produce soluble lignosulfonates. Most of the chemicals produced by this reaction are recovered and used over again.
5. Have students research the paper industry to learn more about paper production.
6. Many substances can be extracted from wood. These include pigments, gums, sugars, fats, waxes, and acids.
7. Methane gas is made by distilling wood in Activity 93.
8. When paper is made industrially by the alkali method, the pulp is heated at 175 °C for 2–3 h. The liquid that is produced is called a *liquor*. It is treated to recover the sodium hydroxide and sodium sulfide originally used.

ANSWERS TO THE QUESTIONS

1. Cellulose is a polysaccharide consisting of long chains of glucose molecules.
2. Hydrogen bonds between molecules of glucose in cellulose and lignin are the primary forces that hold cellulose fibers together.
3. See the Procedure section.

65. Recycling Paper: A Problem for the Chemist

Producing paper from wood is an extensive and expensive process that also uses up our natural resources and pollutes the air and streams with industrial wastes. Chemists are helping to reduce these problems by developing new and better methods for making and reusing some paper products. Reusing paper products is called *recycling*. Today, about 30% of the total paper produced in this country is recycled. In this activity you will carry out the same chemical reactions that the industrial chemist uses to recycle and reuse paper. **This activity takes 2 days.**

Materials

1. Sheet of newspaper.
2. Flour or cornstarch.
3. Wire screen, 8–10 in.^{2} (stapled to a wood frame, if possible).
4. Blender or eggbeater.
5. Beaker, 250 mL.

Procedure

1. Tear half of a sheet of newspaper parallel to one edge. Now make another tear perpendicular to the first. Do you notice any difference? Examine the torn edges. Of what particles is paper made? Continue to tear the paper into small pieces. For best results use the classified section that has only black print on white paper.
2. Place 250 mL (about 1 cup) of water in a large mixing bowl. Add 50 mL of bleach. Add the shredded newspaper and let it sit for a few minutes until the paper becomes soggy.
3. Carefully pour off the water and bleach and rinse the paper twice with water.
4. Add 200 mL of water to the mixing bowl. Using a blender on low speed or a hand-held eggbeater, mix the paper until it forms a thick mush.
5. Add 10 g (about 2 tsp) of flour or cornstarch to the mixing bowl and blend again until the mixture is smooth and soupy.
6. Hold the wire screen over a plastic tray or shallow container. Pour the contents of the mixing bowl onto the wire screen.
7. Spread the mixture on the screen wire as evenly and as thinly as possible with your fingers.
8. Place the wire screen over a tray or pan where it can drain and leave it undisturbed until it is dry. It will take at least 1 day to dry.
9. When the mixture has completely dried, carefully peel it from the wire screen. Describe the paper you have made by the process of recycling. Can you write on it? How does its resistance to tearing compare with the original? How do you account for any differences?

Reaction

Churning the paper in a blender or with an eggbeater increases its surface area. Adding flour or cornstarch puts back some of the polysaccharides and strengthens the paper.

Bleaching helps to remove some of the ink and other impurities.

When the mixture dries, water is removed and hydrogen bonds form between the cellulose fibers. The result is strongly bonded chains of cellulose (polysaccharide) polymer.

Questions

1. How does the paper you recycled compare to better quality paper such as notebook paper?
2. What chemical process occurred when the mixture dried on the wire screen?
3. On the basis of your experience, what problems does one encounter when trying to recycle paper?

Notes for the Teacher

BACKGROUND

This activity gives students experience with an important industrial process with positive environmental implications. Although approximately 50% of all municipal waste is paper, only about 30% of the paper is recycled for reuse. This amount has steadily decreased since the early 1960s when concern for environmental issues peaked. Today, chemists are struggling with problems concerning paper recycling to make it economically feasible to produce a recycled paper that is of high quality.

Solutions and Materials

Have several 8–10-in.^{2} pieces of window screen wire cut at the hardware store. You might want to staple these to wooden frames, or use the wire gauze that you use with tripods and ring stands if it is rustproof.

Either flour or cornstarch will work well.

TEACHING TIPS

1. Recycling paper is difficult because of the many different types of fibers in the variety of paper treated for recycling. Removing inks, glues, and other impurities is difficult, and the process is not economical.
2. Although only about 30% of the paper used in the United States is recycled, the percentage is much higher in some other countries. In West Germany, Great Britain, and Japan, it is at least 50%. Japanese law requires that newspaper consist of at least 50% recycled paper.
3. Most of the recycled paper is used to make boxboard. This type of paper includes cereal boxes and shipping boxes. Only a very small percentage of newspaper is recycled back into newspaper in the United States.
4. Have students examine boxes containing food (e.g., cracker and cereal boxes). A symbol of three small arrows forming a circle indicates that the box is a recycled paper product.
5. Enough paper is recycled in this country each year to fill a volume three times that of the Empire State Building in New York City.
6. Recycled paper is somewhat off-color and weaker than paper made originally from wood pulp.

ANSWERS TO THE QUESTIONS

1. The paper students make in this activity will be slightly colored, rough, and not slick and shiny like notebook paper. To make high-quality paper, the addition of clays, waxes, wet-strength resins, dyes, and permanent inks; extensive bleaching; and better selection of high-quality cellulose fibers are included in the manufacturing process.

2. Cellulose is held together in paper primarily by hydrogen bonds. When the paper is mixed with flour and water and then dried, the hydrogen bonds that were formed with the water then form between flour and cellulose molecules.
3. Some problems encountered might include removing impurities (e.g., ink), producing a white product, drying evenly, speeding up the process, producing a very thin sheet (as in newspaper), and making a strong and durable paper.

66. Measuring Acid Rain

Collecting rainwater regularly to measure acidity can provide important data from widespread sites across the country. Trends and variations can be noted, and changes over time can indicate possible problems to be solved. Normal precipitation is slightly acidic because the normal carbon dioxide in the air dissolves in water to make a slightly acidic solution, about pH 6.5. *Acid rain* is caused by pollutant gases in the air that react with rainwater to make acid. The gases come from burning sulfur in coal and burning nitrogen from the air in the engines of cars. **This activity will take 2 class periods.**

Materials

1. Plastic containers.
2. pH indicator solutions.
3. Wide-range pH paper.
4. Filter paper and funnel.
5. Four or five test tubes.
6. Stirring rod.

Procedure

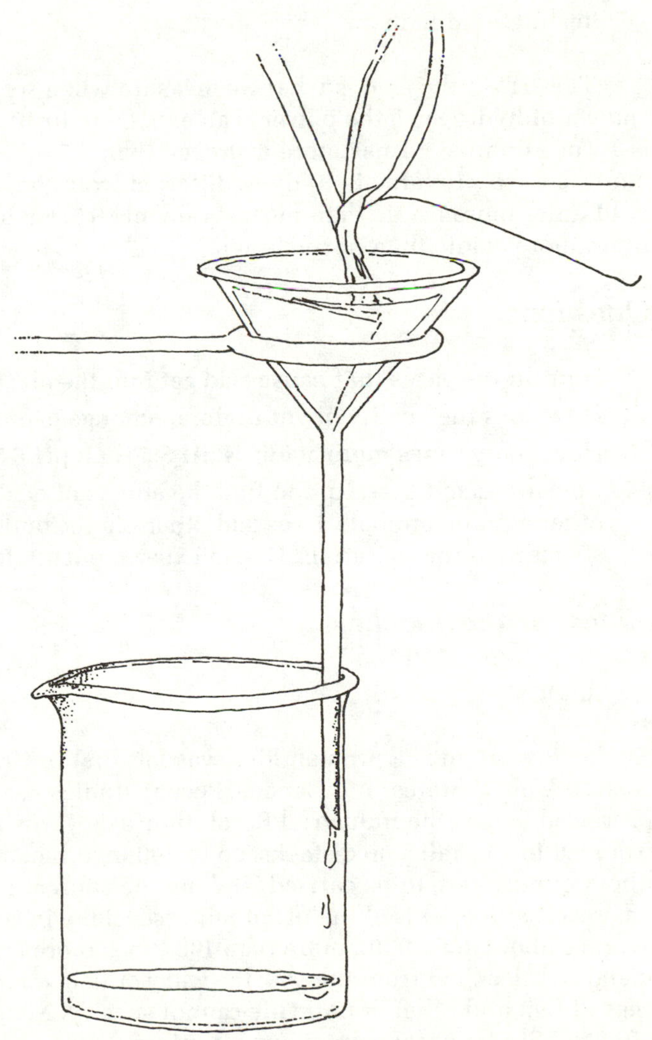

1. Use plastic containers that have been washed and rinsed well.
2. Place the containers in the open rain.
3. Pour the rainwater through filter paper and collect it in another clean plastic container.
4. Dip a clean stirring rod into the rainwater and then touch it to a piece of wide-range pH indicator paper. What is the approximate pH of your rain? pH measures acidity. A value of 7 is neutral. A value less than 7 is acidic. Normal rain is pH 6.5.
5. Divide the rain sample into four or five portions to test with the indicator solution.
6. Choose the indicators that will show color changes in the pH region of your rain.
7. In each of the portions place 1 or 2 drops of a different indicator. Use the charts to find the pH of your rain.
8. Test your rainwater with the most appropriate indicators as often as you can and look for a relationship between acidity and time of year, for example.

Reactions

Many chemical reactions are involved in producing and testing acid rain. Producing acid rain:

$$\underset{\text{sulfur in coal, oil}}{S(s)} + \underset{\text{oxygen in air}}{O_2(g)} \longrightarrow \underset{\text{sulfur dioxide}}{SO_2(g)}$$

$$SO_2(g) + O_2(g) + H_2O(\ell) \longrightarrow \underset{\text{sulfuric acid}}{H_2SO_4(aq)}$$

$$\underset{\text{nitrogen in air}}{N_2(g) + 2O_2(g)} \longrightarrow 2NO_2(g)$$

$$2NO_2(g) + H_2O(1) + O_2(g) \longrightarrow \underset{\text{nitric acid}}{2HNO_3(aq)}$$

Testing Acid Rain:

The "H" in the acid is what we measure when we measure pH. "pH" means "power of hydrogen"; the power is an exponent. In 10^{-1}, the -1 is the power. The pH is 1. For example, -1 has more hydrogen than -7. pH 1 is very acidic, and pH 7 is neutral. Each pH value is 10 times different from the next number. Acid rain, pH 5.5, is 10 times more acidic than normal rain, pH 6.5. Highly acidic rain, pH 4.5, is 100 times more acidic than normal rain.

Questions

1. How do the gases that cause acid get into the air?
2. How does the acidity of your rainfall change over time? (Test it!)
3. How many times more acidic is pH 3.5 than pH 6.5? (See the Reactions section.)
4. Consult recent literature to find the effects of acid rain on soil, water, fish and other aquatic animals, trees and other plants, buildings and other man-made structures, and materials. Design experiments to test these effects for yourself.

Notes for the Teacher

BACKGROUND

Acidic deposition is a problem that was felt first in Great Britain, Scandinavia, the eastern United States, and Canada. Recent studies use the pH of ice in Greenland, produced before the Industrial Revolution, as a yardstick. The pH of this ice ranges from 6.0 to 7.6. Tall smokestacks, up to 300 m, direct sulfur dioxide emissions to the upper atmosphere to be carried far from the source. The acidic precipitation that has resulted has caused the pH of rain, especially in the eastern United States, to average about 4.3 (1980), more than 100 times more acidic than normal rain. Toxic metals are leached from the soil, the waxy cuticle on leaves is damaged, and the eggs of fish and other aquatic life cannot survive. Marble and metal structures are affected. Students can contribute meaningful data about acidity of rain to a pool if they use careful collection and measurement techniques. Although authorities do not agree about the specific sources of specific acidic precipitation, power companies that burn coal are taking steps to remove the sulfur from the coal before it is burned or to remove the SO_2 from the stacks before it enters the atmosphere. Catalytic converters in automobiles reduce the amount of NO_2, which forms HNO_3 (nitric acid) in the atmosphere.

SOLUTIONS

Hydrochloric acid is 6 M. See Appendix 2.

TEACHING TIPS

1. The procedure recommended here for the collection of rain is from *Calibration of Collecting Procedures for the Determination of Precipitation Chemistry* by J. N. Galloway and G. E. Likens, from *Procedures of the First International Symposium on Acid Precipitation and the Forest Ecosystem* (U.S. Department of Agriculture Forest Service Geo-Technical Report NE-23, edited by D. L. Dochinger and P. A. Seliga, Upper Darling, PA, N.E. Forest Experimental Station, 1976). It is also recommended that the plastic containers be cleaned with 6 M HCl followed by 5 rinses of tap water and 5 rinses of distilled water. If you do this, wear gloves, goggles, and a face shield.
2. Galloway and Likens recommend using a pH meter.
3. Contact your local Environmental Protection Agency regional office for information about contributing your students' data to a pool.
4. The indicators that will be most useful for pH 3–7 are the following:
   A. Bromphenol blue: pH 3 (yellow) to pH 4.5 (blue).
   B. Bromcresol green: pH 4 (green) to pH 5.6 (blue).
   C. Methyl red: pH 4.2 (red) to pH 6.3 (yellow).
   D. Bromthymol blue: pH 6 (yellow) to pH 7.6 (blue).
   Check the ranges on other indicators and use any that you have that change color in the acid rain.
5. If students cannot test the water immediately, it should be stored at 4 °C to avoid changes in acidity.
6. Students may want to recreate an acid rain solution such as that which falls in parts of the eastern United States as low as pH 3.1. Use sulfuric acid or nitric acid. The solutions can be used for a variety of tests of the effects of acid rain, including a comparison between sulfuric acid rain and nitric acid rain.
7. Students may also want to research ways to reduce the effects of acid rain.
8. To study the acidity of solutions of carbon dioxide, see Activity 6, "Used Breath: Carbon Dioxide in Exhaled Air", and Activity 103, "Put Fizz in Your Root Beer".

ANSWERS TO THE QUESTIONS

1. From the combustion of coal and oil and the combustion of air in internal combustion engines.
2. Test rainfall during different seasons and over the years or consult tables.
3. 1000 times.
4. Student research.

67. The Gardener's Chemistry Set: Measuring Soil pH

Fruits and vegetables, as well as other plants, thrive at a soil acidity that is in a specific range for each plant. Gardeners test the pH of their soils so that the acidity can be changed to suit the plant. Vegetables generally like slightly acidic soil. A pH of 7 is neutral. Slightly acidic soil is pH 6. You can test the pH of your soil by using a simple technique.

Materials

1. Soil samples, about 100 mL each.
2. Wide-range indicator paper.
3. Plastic containers.

Procedure

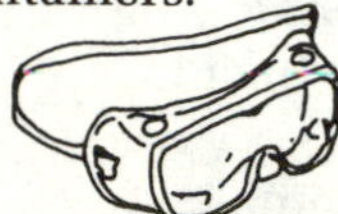

1. Boil some water to use with the soil, about one-fourth of the volume of the soil. Cool the water.
2. Place a strip of pH indicator paper in the bottom of a plastic container, such as a coffee cup, for each sample being tested.
3. Cover the pH paper with a piece of filter paper.
4. Add the soil.
5. Add cooled water until the soil is thoroughly moist but not soaking.
6. Allow the soil to sit for several minutes until the water has moved through the soil to the bottom.
7. Turn the cup over to dump out the soil and papers. Compare the color of the pH paper with the color guide to find the pH of the soil.
8. Test several samples of soil from different spots in the garden. Test a final sample by mixing a portion of all of the samples together before testing.

Reactions

pH is a measure of the balance of hydrogen ions present in a volume of solution. If the solution is neutral, the pH is 7. If it is basic (few hydrogen ions), the pH is between 7 and 14. If it is acidic (extra hydrogen ions), the pH is between 1 and 7. The soil is a complex mixture of substances ranging from bits of plants to minerals. Active fungi will create acids. Many minerals such as limestone will make the soil basic.

Questions

1. What were the pH values of your soil samples?
2. Is a correlation found between the pH of the soil and the types of plants growing in the soil?
3. How would you describe soil of pH 5.0?

Notes for the Teacher

BACKGROUND

Soil has taken hundreds of thousands of years to form as bacteria and fungi have broken down rocks, as heat and water have fragmented rocks, and as roots have gradually made their way through the tiny cracks. Soil is about one-fourth air, one-fourth water, and one-third to one-half minerals; the rest is organic matter. This balance helps to determine the pH of the soil. The amount of acid in the soil determines which nutrients are available to the plants. Too little acid causes iron and other metal nutrients to be unable to dissolve because they form insoluble carbonates or hydroxides under this condition. Too much acid causes the nutrients to dissolve so well that they are carried away or leached too deep for the plant to reach, or they become so concentrated in the soil that they are toxic.

TEACHING TIPS

1. The water is boiled to remove carbon dioxide, which makes the water acidic (pH about 6.5). If you have hard water, you should use distilled water for this activity. All tap water has some impurities dissolved in it. Hard water has calcium and magnesium salts dissolved in it, particularly carbonates, which can cause the pH of the water to be slightly basic.
2. Plastic containers are preferred but are not necessary. Some metals that change soil pH can be leached out of some glasses.
3. Litmus paper can be used. Use a piece of blue and a piece of pink paper for each sample.
4. See Activity 66, "Measuring Acid Rain", for more practice with pH.
5. You might have students sharpen their powers of observation by finding correlations between the types of plants and the types of soils.

   Plants that prefer acidic soil (pH 4–6): blackberries, blueberries, marigolds, oak trees, rhododendrons, peanuts, lilies of the valley, watermelons, spruce trees, and sweet potatoes.

   Plants that prefer basic soil (pH 7.5–8.5): beans, asparagus, beets, lettuce, peas, onions, irises, nasturtium, phlox, carnations, cantaloupes, and cucumbers.

   Many other plants prefer neutral soil.
6. Soil pH can be changed. If the soil is too acidic, add limestone ($CaCO_3$). The carbonate ion (CO_3^{2-}) reacts with acid to make carbon dioxide. Wood ashes, a mixture of basic salts, may be added.
7. If the soil is too basic, add grass clippings, decaying leaves, peat moss, pine needles, or nut shells. These produce acid as they decompose.

ANSWERS TO THE QUESTIONS

1. Answers will vary.
2. A correlation may be found between moss and fungi and acidic soil.
3. Very acidic.

Biochemistry: Chemistry of Living Things

68. Enzymes in Cells

Why does hydrogen peroxide decompose and produce oxygen bubbles when exposed to broken cells such as on a cut in your skin? The decomposition of hydrogen peroxide is caused by a specialized enzyme molecule, *catalase*, which is found in cells, in your blood, and in turnips and potatoes. You will investigate the presence of catalase in potatoes. You will also study the effect of temperature, as in cooking, on the activity of enzymes.

Materials

1. Five test tubes, 13 × 100 mm.
2. Fresh 3% hydrogen peroxide (H_2O_2), 20–25 mL.
3. Sample of ground potato pulp, about 10 mL.
4. Metric ruler.
5. Four water baths: ice bath (approximately 0 °C), room-temperature bath (20–25 °C), body-temperature bath (about 37–40 °C), and high-temperature bath (70–80 °C).
6. Clock or other timing device.

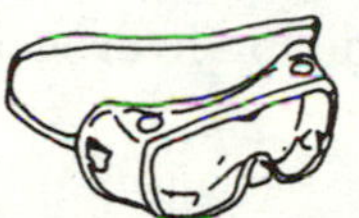

Procedure

1. Place potato pulp into one test tube to a depth of about 2 cm.
2. Add an additional 3 cm of hydrogen peroxide. Stir.
3. Observe the formation of oxygen gas bubbles by the action of the potato enzyme on the hydrogen peroxide.
4. Measure the height of the foam after 1 min. Measure from the top of the liquid to the top of the foam. Some pulp may be caught in the foam.

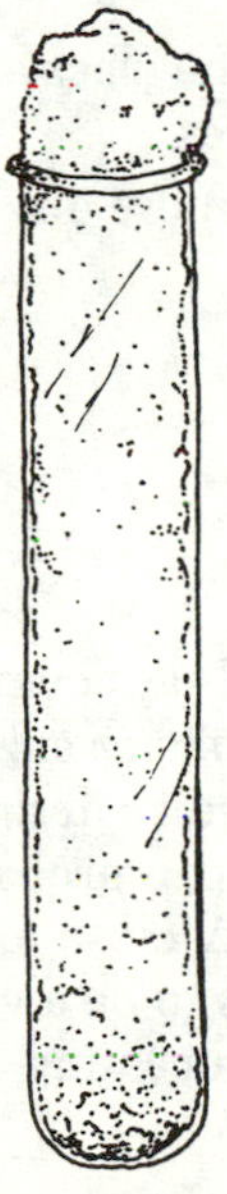

5. Measure the foam height each minute for a total of 5 min.
6. Construct a graph of foam height on the *y* axis versus time on the *x* axis. Label the graph as GRAPH 1.
7. Label four more test tubes: 1, 2, 3, and 4.
8. Place potato pulp about 2 cm deep in each test tube.

9. Place each tube at a constant temperature for several minutes:
   1: ice bath (approximately 0 °C).
   2: room-temperature bath (20–25 °C).
   3: body-temperature bath (37–40 °C).
   4: high-temperature bath (70–80 °C).
10. Remove the tubes from the baths and add the same amount of hydrogen peroxide to each, about 3 cm up the test tube. Put the tubes back into the baths.
11. Compare the rates of oxygen production by measuring the foam height in each tube after 5 min.
12. Construct a graph of foam height on the *y* axis versus temperature on the *x* axis. Label the graph as GRAPH 2.
13. Design an experiment to investigate the activity of catalase from other animal and plant products such as turnips, apples, and liver.
14. Design an experiment to test the activity of catalase as acidity is varied.

Reaction

The enzyme, catalase, acts as a catalyst to cause the decomposition of hydrogen peroxide. One catalase molecule can act on up to 5 million hydrogen peroxide molecules in 1 min.

$$2H_2O_2 \xrightarrow[\text{catalase}]{} O_2(g) + 2H_2O$$

Questions

1. Describe what happens in the first 5 min after potato pulp at room temperature is added to hydrogen peroxide.
2. Examine the results of other members of the class and compare the rates of oxygen production at different temperatures.
3. Why do you think it is important to decompose any hydrogen peroxide produced in cells?

Notes for the Teacher

BACKGROUND

The chemical reactions that occur in living cells are complex and highly controlled. Practically all of them involve the use of catalysts or enzymes. These are large protein molecules that have a specialized structure designed to work with a particular reaction. Thousands of different enzymes can probably be found in any cell. Chemists have been able to identify and classify close to 2000 altogether. They have found the isolation and description of enzymes to be a very active field of research. Enzymes are also useful industrially in food processing and in detergent manufacturing.

SOLUTION

Hydrogen peroxide (3%) is the drugstore variety. It must be fresh.

TEACHING TIPS

1. Red-skinned potatoes and Irish potatoes are the richest in catalase. One peeled potato with a diameter of about 8 cm will provide enough pulp for 6–8 student teams. The best way to produce the pulp is with a blender. Place the cut-up potato in a blender with about ½ cup of water. The pulp can be used directly or can be filtered. Coffee filters or cheesecloth works best. Use the filtrate (the solution that comes through the filter).
2. The catalase is apparently a stable molecule. Even after the pulp has remained uncovered at room temperature for several hours, the activity is not diminished.
3. Typical results:

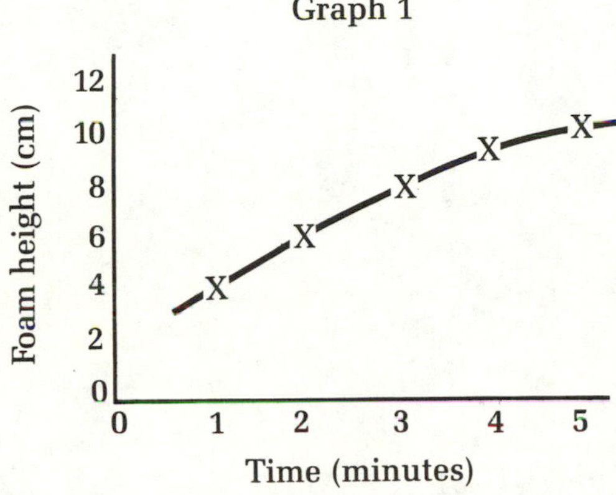

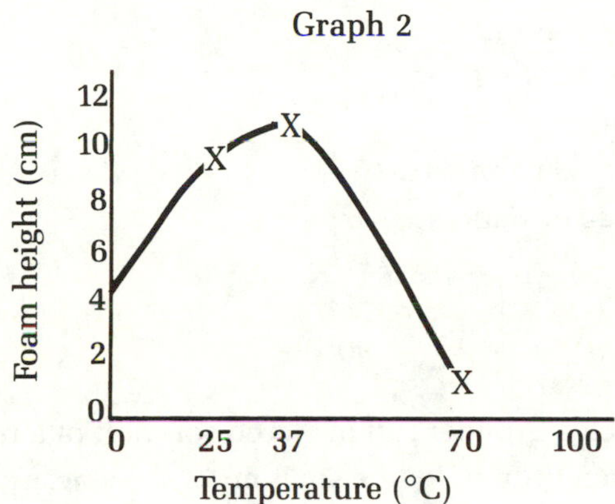

4. This experiment offers students the opportunity to design some procedures to test the effects of variables. You could encourage students to do this activity at home because most of what is required is usually found in the home. Each student will need several test tubes. If the students keep the test tubes at home, several other experiments can be conducted at home throughout the year.
5. You may wish to have a few central water baths for the two elevated temperatures.
6. As students become familiar with controlling variables, suggest tests of catalase activity of other foods such as turnips and other tubers.

ANSWERS TO THE QUESTIONS

1. Oxygen-filled bubbles form, and the foam increases until it rises up out of the tube.
2. Combining class results will usually produce the clearest picture under the conditions in your laboratory.
3. Hydrogen peroxide produced in cells is toxic.

69. Enzymes in Saliva

When foods containing starches are chewed in the mouth, they are mixed with saliva. Saliva contains an enzyme, amylase, that helps degrade starch into small sugar molecules. Starch turns purple when added to iodine, but the small sugar molecules do not. In this activity you will add saliva to starch to cause the starch to break down and prevent it from reacting with iodine.

Materials

1. Soluble starch solution.
2. Iodine solution.
3. Buffer solution.
4. Beakers, 250 mL.
5. Droppers.
6. Graduated cylinders.
7. Dextrose or maltose.

Procedure

1. Test for starch.
   A. Place a dropper full of starch solution in a test tube.
   B. Add a drop of iodine solution to the starch.
   C. Notice the deep purple. This test confirms the presence of starch.
   D. Record your observations.
   E. Add a small amount of the sugar, dextrose or maltose, to 10 mL of water.
   F. Add a drop of iodine solution. What happens?
   G. The sugars that result when starch is broken down do not turn purple in the presence of iodine.
2. Test your saliva for amylase.
   A. Make a saliva solution: Add 100 mL of buffer solution to a beaker. Add about 2 mL of your own saliva to the beaker and stir until it dissolves. Label this "saliva solution", including your own name.
   B. Add 10 drops of iodine solution to the beaker.
   C. Add 20 drops of starch solution to the beaker and stir. The solution should be light blue. Record the time.
   D. Let the solution stand for a few minutes.
   E. What happens? Record the time when the blue disappears.
   F. Compare your amylase activity with that of others in the class.

Reactions

1. Starch reacts with the iodine, as I_3^-, to produce the blue starch-iodine complex:

$$I_2 + I^- \longrightarrow I_3^-$$

$$I_3^- + \text{starch} \longrightarrow \underset{\text{blue}}{\text{starch-}I_3^-\text{ complex}}$$

2. When the enzyme, amylase, in your saliva reacts with the starch, it causes the starch to break down into smaller sugar molecules that do not react with iodine. When the starch was all degraded, none was present to react with the iodine, and the blue disappeared.

Questions

1. What foods do you think contain starch? How can you test for starch?
2. Enzymes such as amylase are destroyed by heat. How could you show this destruction experimentally?
3. Why does it take several minutes for the blue to fade? Why does it not disappear immediately?

Notes for the Teacher

BACKGROUND

This activity will allow students to study a simple and readily available enzyme, salivary amylase. Amylase acts to partially degrade starch into simple disaccharide sugars (maltose), smaller polysaccharides (dextrin), and simple sugars (glucose). None of these products give a positive iodine test, but starch does. Students will prepare an enzyme solution and test its ability to break down starch.

SOLUTIONS

1. Iodine solution: Dissolve 2.5 g of potassium iodide (KI) in 500 mL of distilled water. Add 1.0 g of iodine (I_2) and stir until all particles are dissolved. Place in dropper bottles.
2. Starch solution: Make a paste of about 5 g of soluble starch in 15 mL of hot water. Dissolve this paste in about 100 mL of boiling water. Cool the solution and dilute it to 1 L.
3. Buffer solution: To make 3 L, dissolve 30 g of potassium dihydrogen phosphate (KH_2PO_4) in about 500 mL of water. Add 2.5 g of sodium hydroxide (NaOH). Dilute to 3 L. Avoid contact with sodium hydroxide. It is corrosive. Plastic 1-gal (3.78-L) milk jugs are convenient for the final mixing, dispensing, and storing of buffer solutions.

TEACHING TIPS

1. The saliva solution is a buffer solution with a pH of about 6.5. Enzymes are sensitive to changes in pH. This enzyme, amylase, has optimum activity at room temperature at pH 6.5.
2. Amylase gets its name from *amylose*, one of the two long-chain polymers that make up starch. The other chain, which is branched, is called *amylopectin*. Amylose forms a coiled structure. Iodine, probably as the species I_5^-, fits inside this structure to form the blue complex.

starch–iodine complex 

3. The iodine solution is made in KI because iodine is soluble in a KI solution but not in water alone.

4. Some saliva has a very low concentration of amylase. Thus, some students may need to use more saliva in their solutions.
5. The pancreas also produces amylase. This substance is called pancreatic amylase and is part of pancreatic juice. It is responsible primarily for the digestion of starch in the small intestine.
6. See Activities 26, 48, 76, and 87 for additional experiments with starch and iodine.
7. You may see some individual variation in reaction times within the class. Stress the importance of controlling all other variables. You may wish to construct a normal distribution curve with your students' results.

ANSWERS TO THE QUESTIONS

1. Potatoes and bread. Test some with iodine solution.
2. Repeat the experiment after heating the saliva solution.
3. All of the starch must react. The enzyme molecules are used over and over again.

70. Chemistry of Bruised Fruit

When fruit and some vegetable cells are broken and exposed to the air, certain enzymes react with the oxygen in the air. These enzymes, called *oxidative enzymes*, begin to digest the cell, and this process produces the brown color seen in cut apples and bananas, for example. Vitamin C (ascorbic acid) in fruit normally keeps the oxygen from reacting with the enzyme. Vitamin C is a naturally occurring reducing agent. You will investigate the interaction of oxygen gas and ascorbic acid by studying the colors of cut fruit.

Materials

1. Fresh fruit, preferably apples.
2. Vitamin C solution, 100 mg/100 mL solution.
3. Fruit juice such as orange or tomato juice.
4. Boiled water.
5. Vinegar (acetic acid, 5%).
6. Six small test tubes.

Procedure

1. Prepare the following six test tubes in a rack. Label each tube.
   Tube 1: Open to the air.
   Tube 2: Half-filled with tap water.
   Tube 3: Filled to the top with water that has been previously boiled and stoppered.
   Tube 4: Half-filled with vitamin C solution.
   Tube 5: Half-filled with fresh fruit juice.
   Tube 6: Half-filled with vinegar, a weak acid.
2. Cut an unpeeled apple into six pieces about 5 cm long and about 1 cm on each side.
3. Bruise each piece with your thumb and immediately place one piece in each test tube.
4. Be sure to put a stopper in tube 3.
5. After 20 min, examine each apple slice.
6. Using the amount of discoloration in tube 1 as a standard, record the relative amount of discoloration in each tube.

Reactions

The vitamin C molecule is easily oxidized by the oxygen in the air or by the enzyme. If vitamin C is present, the enzyme cannot be active, and the fruit will remain white. However, the vitamin C molecule is changed when it reacts with oxygen and can no longer be an active vitamin.

Questions

1. What variable is being tested in each test tube?
2. Arrange the tubes in order from the most darkened to the least darkened.
3. What variables seem to cause the darkening of fruit?
4. What variables are associated with nondarkened fruit?
5. Why is lemon, lime, or orange juice squeezed over a fruit salad containing apples, bananas, or pears?

Notes for the Teacher

BACKGROUND

The enzyme that causes darkening in fruit is one of many found in the cells of both plants and animals. This enzyme is believed to cause the darkening of skin by participating in the production of *melanin*, the pigment in skin that gives it its dark color. Ascorbic acid (vitamin C) is an active reducing agent in cells. It reduces molecules that have been oxidized. It reacts with oxidative molecules to protect other molecules from oxidation. The mechanisms of vitamin C activity, however, are not known. Although vitamin C is required for normal functioning of cells, the human body cannot synthesize it, and it must be replenished in the diet each day.

SOLUTIONS

1. A 100 mg (0.100 g) tablet of vitamin C should be enough to prepare 100 mL of vitamin C solution. Prepare just before using. Vitamin C is so easily oxidized that a measurable amount is lost in only a few minutes. Each team will need about 5–10 mL.
2. Have a filled covered jar of boiled tap water prepared for use by the class. Boiling water reduces the solubility of oxygen in water and effectively removes the oxygen. Air is about 20% oxygen.

TEACHING TIPS

1. This activity can serve many purposes. It provides nonquantitative practice with experimental design and variable control. It introduces the ideas of enzymes, oxidation, and organic acids.
2. Another experiment might test the darkening behavior in vitamin C solutions of varying concentrations.
3. Students might like to do the quantitative tests for vitamin C described in Activity 80.
4. Students should not eat the fruit.

ANSWERS TO THE QUESTIONS

1. Tube 1: Control.
   Tube 2: Effect of water with oxygen.
   Tube 3: Effect of water without oxygen.
   Tube 4: Effect of vitamin C with water with oxygen.
   Tube 5: Effect of vitamin C in fruit juice (including water with oxygen).
   Tube 6: Effect of an acid (including water with oxygen).
2. Probably tubes 1, 2, 6, 3, 5, and 4 is one observed sequence.
3. Presence of oxygen in water or air.
4. Absence of oxygen or presence of vitamin C.
5. Vitamin C in juices prevents the darkening of fruit.

71. Calcium and the Rubber Chicken

The body contains more calcium than any other mineral, and 99% of this calcium is in bone. Calcium makes bones strong, hard, and durable. In this activity you will chemically remove the calcium ions from the long bone of a chicken leg. The decalcified bone will not be firm and strong. It will be so flexible and rubbery that you can tie it into a knot! **This activity will take 1 week.**

Materials

1. Chicken leg.
2. Vinegar (acetic acid, 5%).
3. Jar.
4. Pan.

Procedure

1. Boil a chicken leg in a pan of water until all of the meat is removed or loosened. Pick all of the meat from the bone.
2. Place the bone in a large jar and completely cover it with vinegar. Lightly cap the jar or cover it with plastic wrap.
3. After 2 days, pour off the vinegar and examine the chicken bone. How did it change? Add fresh vinegar and allow the bone to soak for several more days.

Reaction

Most of the calcium in bone is in the form of insoluble hydroxyapatite [$Ca_5(PO_4)_3(OH)$]. Although this structure is impressive, it essentially consists of calcium as Ca^{2+} ions. These calcium ions will react with vinegar to produce soluble calcium acetate. Thus, the calcium will be removed from the structure in the bone:

$$Ca^{2+} + 2CH_3COOH(aq) \longrightarrow Ca^{2+}(CH_3COO^-)_2(aq) + 2H^+(aq)$$

Without calcium, the rigid structural component of the bone is lost, and the bone becomes elastic rather than unbendable.

Questions

1. What structure in bone makes calcium insoluble?
2. How does vinegar remove calcium from bone?
3. What are other minerals that are important in bone formation? (Go to the library!)

Notes for the Teacher

BACKGROUND

Calcium is one of the major minerals of the body. It is essential for the proper development of bone. The calcium in bone exists primarily combined with phosphorus in a stable, insoluble, complex structure called *hydroxyapatite*. In this activity we will soak the bone in acid. This step will lead to the formation of the soluble compound, calcium acetate. The calcium-deficient chicken bone will lose its strength and rigidity and become limp and rubbery.

TEACHING TIPS

1. Bone is a living tissue that has a collagenous protein matrix impregnated with mineral salts, especially calcium phosphate. When the calcium is removed, the bone loses its rigid structure.
2. You can do a similar activity with a chicken egg. The shell consists primarily of calcium compounds. When soaked in vinegar, the egg also becomes rubbery. The eggshell may completely dissolve, and only the membrane supporting the egg will be left.
3. Have students research the role of calcium in the body and its regulation.
4. Try to identify the calcium in solution after the 2 days. Calcium gives an orange-red flame test. Squirt some of the solution into a blue Bunsen burner flame, or straighten a paper clip, dip it into the solution, and put it into a flame. You might also try to precipitate calcium from the solution by adding ammonium oxalate. Heat the solution in a hot water bath for about 3 min. Then add a few drops of 0.5 M ammonium oxalate [$(NH_4)_2C_2O_4$] solution and mix well.

   Insoluble calcium oxalate will precipitate as a white solid. (You might also separate this solid and try the flame test with it after redissolving with concentrated hydrochloric acid.)

$$Ca^{2+}(aq) + C_2O_4^{2-}(s) \longrightarrow CaC_2O_4(s)$$

5. Students should not eat the meat if it is prepared in the laboratory.

ANSWERS TO THE QUESTIONS

1. Hydroxyapatite [$Ca_5(PO_4)_3(OH)$].
2. It produces the soluble salt, calcium acetate.
3. Mainly phosphorus and magnesium. Other trace minerals are important to some extent.

72. Detecting Nitrogen in Hair

Hair consists primarily of a protein called *keratin*. This protein is also found in fingernails, feathers, horn, wool, and skin. All proteins are polymers consisting of smaller units called *amino acids*, and all amino acids contain the element nitrogen. Keratin also contains the element sulfur, which accounts for the strong odor when these proteins are burned. We can treat keratins (and other proteins as well) with a strong base to cause them to break down. When they do, the nitrogen is released as ammonia. We will detect the ammonia because of its familiar odor and its reaction with litmus indicator.

Materials

1. Samples of hair.
2. Calcium oxide (CaO).
3. Litmus paper (red).
4. Dilute ammonia solution (NH_3) (aq).
5. Test tubes.
6. Burner.

Procedure

1. Using a stirring rod, loosely pack about ½ in. of hair in the bottom of a test tube.
2. Add 1.0 g of calcium oxide to the tube.
3. Add water to the tube until the hair is just covered.
4. Gently heat the tube with a burner until it begins to boil.
5. Moisten a piece of red litmus paper and hold it over the mouth of the test tube.

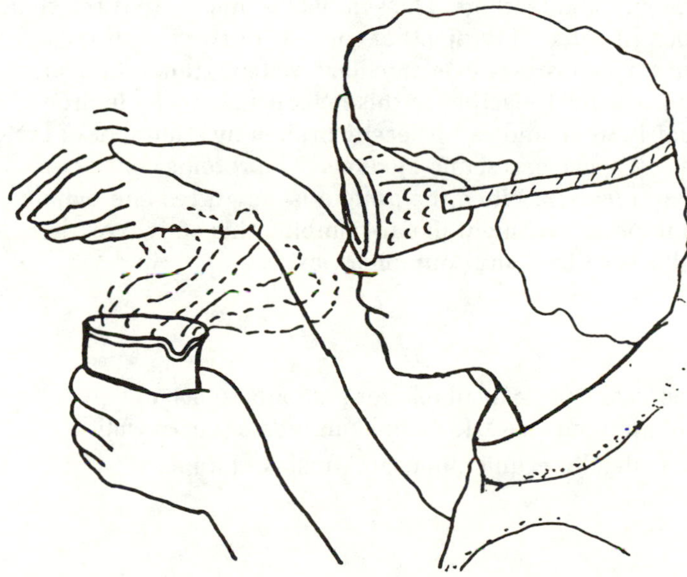

6. When the paper turns blue, carefully waft some of the gas from the tube toward your nose. NEVER SNIFF ANYTHING DIRECTLY FROM A TUBE OR BOTTLE. Can you identify the gas as ammonia? Try the smell test with a bottle of dilute ammonia. How does its smell compare with the odor from the test tube?

Reactions

Certain factors and substances will break down, or *denature*, proteins. These factors include heat, alcohol, acids, bases, and detergents. In this activity we made calcium hydroxide, a very strong base, by reacting calcium oxide and water:

$$CaO(s) + H_2O \longrightarrow \underset{\text{calcium hydroxide}}{Ca(OH)_2(aq)}$$

The reaction is *exothermic*, and the two conditions (base and heat) caused the protein, keratin, to break down into amino acids. The nitrogen in the amino acids was then released as ammonia:

$$\text{protein} \longrightarrow \text{amino acid} \longrightarrow \text{ammonia } (NH_3)$$

Questions

1. Did the litmus paper change color? Why?
2. How did you detect the presence of nitrogen?
3. Would this experiment work with feathers and similar structural proteins (e.g., fingernails)? (Try them!)

Notes for the Teacher

BACKGROUND

The protein molecule is a very complex structure. α-Keratin is a structural protein found in wool, skin, hair, nails, and feathers. The symbol "α" means that the structure of this protein is much like that of a rope that consists of twisted strands of smaller rope, and these in turn consist of even smaller twisted strands, and so on. The individual amino acids are held together in this helix primarily by hydrogen bonds. These bonds are fairly weak and can be easily broken by a number of factors and substances such as heat, acids, bases, and alcohols. All proteins consist of amino acids, which contain nitrogen. When the protein is denatured completely, nitrogen is released as ammonia. You might like to combine this activity with Activity 57, "Smelling Salts, Fertilizer, and Ammonia Gas".

MATERIALS

1. Have a bottle of aqueous ammonia available for students to use to detect the characteristic smell of ammonia gas. Household ammonia works well.
2. Calcium oxide is also called lime, quicklime, and unslaked lime.

TEACHING TIPS

1. Demonstrate the proper way to smell a gas from a tube or bottle. Hold the container 6–8 in. from the nose and waft the odor toward the nose with a gentle hand motion over the tube.
2. You can perform the same experiment with wool, feathers, or nail clippings.
3. An egg is denatured when it is heated. The albumin (protein) in the egg white becomes white and stiff.
4. Ask students to think of other examples of protein denaturation (e.g., the peeling of skin as a result of sunburn and the killing of bacteria with alcohol).

5. You must be very careful when storing calcium oxide. A large quantity, if mixed with water, can generate enough heat to ignite paper or even wood.
6. See Activity 57 for a discussion of ammonia.

ANSWERS TO THE QUESTIONS

1. The basic ammonia produced in the reaction turned the red litmus paper blue.
2. Nitrogen is part of the ammonia molecule (NH_3). Ammonia compound can be detected by its characteristic odor.
3. Yes, all of these proteins produce ammonia when denatured.

73. Fermentation and Apple Cider

Fresh fruits such as apples contain the ingredients necessary for the process of fermentation: yeast, sugar and other carbohydrates, and water. When oxygen in the fruit mixture is lacking, the enzymes in the yeast produce ethyl alcohol. In this procedure you can control the amount of oxygen and cause the yeast enzymes to produce ethyl alcohol. You may find that if too much oxygen is present, another enzyme reaction will produce acetic acid instead, and the result will be vinegar. **This activity will take 2–3 days.**

Materials

1. Ripe apples.
2. Knife.
3. Beaker.
4. Mortar and pestle or blender.
5. Muslin or other cotton cloth.
6. Flask or bottle with one-hole stopper to fit, 125 mL.
7. Small length of glass tubing to fit into stopper hole.
8. Rubber tubing, 20–30 cm long.

Procedure

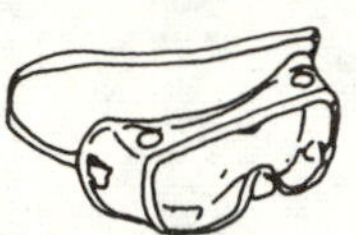

1. Cut the apples, with skin left on, into small pieces. Crush the pieces with the mortar and pestle or with the food blender.
2. Put the pieces into a square of cloth, twist the ends closed, and continue to twist to force out as much juice as possible.
3. Pour the apple juice into the flask to within 4 cm of the top.
4. Assemble the rubber stopper with the glass piece attached to the rubber tubing.

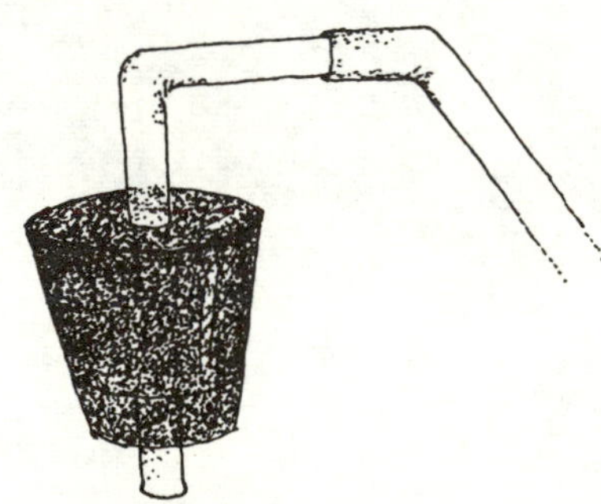

5. Put the stopper into the flask and the end of the rubber tubing into the beaker filled with water.
6. Let the apple juice stand for 2–3 days. Continue to check for evidence of a reaction.
7. How can you determine whether yeast enzymes have produced ethyl alcohol or vinegar (acetic acid)? (You could test the acidity with pH paper.)

Reaction

The reaction to produce ethyl alcohol is the following:

$$\underset{\text{simple sugar}}{C_6H_{12}O_6(aq)} \xrightarrow{\text{enzymes}} \underset{\text{ethyl alcohol}}{2C_2H_5OH(\ell)} + 2CO_2(g)$$

Questions

1. Describe the cider product formed by your fermentation process.
2. What evidence indicated that a chemical reaction was taking place?
3. Why do you think the skin was left on the apples?

Notes for the Teacher

BACKGROUND

Fermentation is a natural process that occurs on a large scale. It is the oldest known chemical process used by humans and has been traced back at least 5000 years. Fruits, raisins, potatoes, corn, rice, table sugar, molasses, grapes, and certain grains will all ferment. When the yeast enzymes produce an alcohol solution that is 12% alcohol, they become inhibited. A higher concentration of alcohol must be obtained by distillation. Beers are produced by fermentation of carbohydrates such as barley and hops to an alcohol content of 3–10%. Wines are produced by fermentation of fruit juices to an alcohol content of 10–14%. "Spirits" are distilled from various mashes of fermented carbohydrates. Fermentation is also the process that produces the carbon dioxide in bread dough, which causes the dough to rise.

TEACHING TIPS

1. Students might like to see the yeast cells under the microscope. A slide could be prepared from the sediment in the flask.
2. Students may need to pool their samples of apple juice so that the level of juice is within 4 cm of the top of a flask. Excluding as much air as possible is important.
3. Use caution when inserting the piece of glass tubing into the stopper. Moisten with a little water or glycerin.
4. The rubber tubing is placed in the beaker of water to allow gases to escape from the "cider mill" without introducing air. Check to see that the tube is not clogged because pressure may build up in the flask.
5. Students should observe the formation of the carbon dioxide. You may want to demonstrate the test for carbon dioxide. Place the rubber tubing in a beaker of limewater (calcium hydroxide solution). The limewater will turn milky to indicate the presence of carbon dioxide. The reaction produces calcium carbonate. See Activity 6, "Used Breath: Carbon Dioxide in Exhaled Air".
6. Students can test for some of the products by determining pH with pH paper and by performing a distillation. Place the solution in a larger flask, use a similar stopper and tubing assembly, heat to the boiling point, and collect the distillate that condenses in a test tube kept cool by a beaker of water. If you use a two-hole stopper, a thermometer can be inserted in one hole so that the temperature can be monitored. See Activity 92, "Distillation of Vinegar".
7. The distillate could be tested for flammability. In the 1600s, this test was done on a small pile of gunpowder. The gunpowder would ignite after the alcohol had burned off if the alcohol was at least 50% pure. This result was "proof" of the quality of the alcohol, usually whiskey. Fifty percent alcohol is 100 proof.
8. Students must not drink the cider.
9. Students may want to investigate the history of fermentation.

ANSWERS TO THE QUESTIONS

1. The cider should be clear and light brown. Sediment of yeast may be visible at the bottom of the flask.
2. Gas bubbles indicate a chemical reaction.
3. A lot of yeast is present on the skin.

74. Pectin in Fruit Juices

Do you want to turn your applesauce into apple juice? Apple cells make *pectin*, which is a large polysaccharide molecule that keeps particles from settling out in the fruit cells. If the large molecule of pectin can decompose into smaller units, then the material in cells will settle out, and juice will be released from the fruit cells. You will investigate the activity of an enzyme, *pectinase*, as it helps to change pectin molecules from large colloidal molecules to small individual sugar molecules.

Materials

1. Applesauce, 250 mL.
2. Pectinase enzyme.
3. Droppers.
4. 10 plastic cups or beakers.
5. Graduated cylinder, 100 mL.
6. Coffee filters or filter paper.
7. Funnel.

Procedure

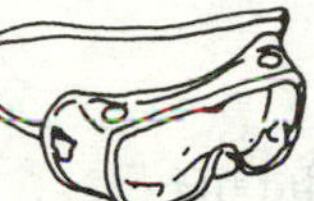

1. Label pairs of cups or beakers with the amount of pectinase enzyme that will be used in each test: 0 mL, 0.25 mL (5 drops), 0.50 mL (10 drops), 0.75 mL (15 drops), and 1.0 mL (20 drops).
2. Add an equal amount of applesauce to one of each pair of cups, about 50 mL. The exact amount is not important as long as you know the amount.
3. Add the drops of pectinase to the cups of applesauce according to the labels.
4. Stir and allow the applesauce to stand for 10 min.
5. Using coffee filters or filter paper and funnels, filter the juice into the empty cups.
6. Use a clean graduated cylinder or measuring cup to find the volume of juice produced in each case.
7. Plot a graph of the pectinase concentration (in drops or milliliters) on the *x* axis versus the volume of juice on the *y* axis.

Reaction

Pectin is a *mixed polysaccharide*, a long-chain polymer composed of smaller sugar units. When in a plant cell, which concentrates sugar molecules, the pectin holds together a suspension of these sugars, other substances, and the pectin molecules. Pectinase breaks the bonds in the long-chain polymer of pectin. The juices are released from the apple cells.

Questions

1. How did the amount of pectinase affect the amount of juice that resulted?
2. What practical application does the addition of pectinase to pulp have?
3. Why do you think fruits do not readily release their juices?

Notes for the Teacher

BACKGROUND

Pectin is a molecule that belongs to a class known as *polysaccharides*, which are large complex molecules made up of repeating simple sugar units. A single molecule may have tens of thousands of atoms, mostly carbon, oxygen, and hydrogen, because polysaccharides are carbohydrates. Polysaccharides constitute 50–90% of the dry mass of plants. Other common polysaccharides are starch, cellulose, chitin (insect and crab shells), and agar. Enzymes are responsible for the rate and control of chemical reactions in living cells and can be used under the right conditions to catalyze these same reactions in laboratory glassware or in plastic cups. Pectinase is used in the commercial juice and wine industry to optimize the amount of juice produced.

TEACHING TIPS

1. Pectinase is available from any biological supply house.
2. Students should not eat the fruit or drink the juice.
3. Some of your students may want to try other fruit pulps.
4. In addition to testing the amount of pectinase as a variable, try varying the pH or temperature while keeping the number of drops constant.
5. Some students might like to do library research on the fluctuation of the free-flowing "sol" state and gel state in the cytoplasm of living cells.

ANSWERS TO THE QUESTIONS

1. Answers will vary. However, depending on the type of applesauce, you may expect that greater amounts of enzyme will lead to more juice being produced in 10 min.
2. In industry, pectinase is used to increase the yield of juice.
3. In living cells, pectinase is kept isolated from the cytoplasm.

75. Tests for Protein

Is protein present in your food? Find out by performing the biuret test. Protein is an important class of molecule needed in your daily diet because much of each body cell is made of proteins. The antibodies that fight disease are proteins, as are your hormones. The enzymes that control reactions in your cells are proteins. The word "protein" comes from the Greek word "proteios", which means "of prime importance". Your cells and other proteins are built from the proteins you eat. These dietary proteins are broken down into amino acids during digestion of your food. The amino acids are used to make body proteins.

Materials

1. Food samples (e.g., egg white and luncheon meats).
2. Sodium hydroxide (NaOH) solution, 6 M.
3. Copper sulfate ($CuSO_4$) solution, 0.1%.
4. Several droppers.
5. Test tubes or small plastic cups.
6. Stirring rod.

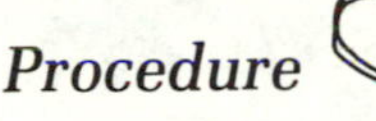

Procedure

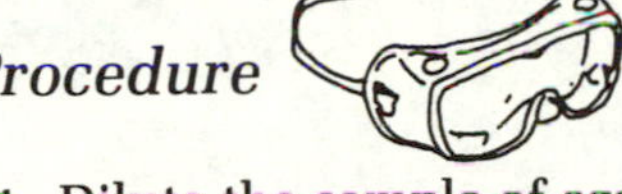

1. Dilute the sample of egg white with water and place 2 mL of it in a test tube or cup.
2. Add 4 full droppers (about 2 mL) of NaOH solution. BE CAREFUL! Wear goggles, gloves, and a face shield for this step. Gently mix the solution with a stirring rod.
3. Add a few drops of 0.1% $CuSO_4$ solution.
4. Mix thoroughly. Add more drops of $CuSO_4$, up to about 10 drops, for a more intense color change.
5. Compare the reactions of various foods. Grind solid samples, like luncheon meats, in a small amount of water for testing.
6. When you have some positive tests for protein, try diluting the sample by one-half and one-fourth. Test again and compare the intensities of colors of the dilutions.

Reaction

In a solution of base (NaOH or other base), copper sulfate forms a complex with the chemical bond that holds the parts (amino acids) of proteins together. The complex, called "biuret", is purple. The chemical bond holding amino acids together is called the *amide bond* or *peptide bond.*

Questions

1. What substances tested contain detectable protein?
2. Compare the results from each food.
3. How do the color intensities before dilution compare with those after dilution?
4. Why do you need protein molecules in your food?

Notes for the Teacher

BACKGROUND

Proteins are complex organic molecules synthesized in living cells for the functions listed on the student page. A protein molecule is a combination of several amino acids, of which 20 kinds are common. Hemoglobin, for instance, has chains of 574 amino acid building blocks. The possible number of combinations and sequences of amino acids is staggering. Protein must be present in food because your body cannot make eight kinds of amino acids. Therefore, you need to get these amino acids from the protein in your food. The test for protein shows a positive result whenever one amino acid is linked to another one.

SOLUTIONS

1. Sodium hydroxide (NaOH) is 6 M. Dissolve 240 g of NaOH per liter of solution. Wear gloves, goggles, and a face shield.
2. Copper sulfate ($CuSO_4 \cdot 5H_2O$) is very dilute. Dissolve 0.1 g of $CuSO_4 \cdot 5H_2O$ per 100 mL of solution.

TEACHING TIPS

SAFETY NOTE: Sodium hydroxide solution is caustic. Use care. Have students add the sodium hydroxide at a station equipped with a face shield.

1. Scientists use several methods to test for proteins. This test, the biuret test, is the most general because it tests for the presence of the amide or peptide bond.
2. Samples may be brought from home. You might try testing wool (protein) and nonprotein cotton (cellulose) also.
3. The positive test produces a blue-violet color.
4. The intensity varies with the concentration of protein.
5. To confirm the biuret test you may do the xanthoproteic test: Add 1 mL of concentrated nitric acid (HNO_3) to 2 mL of sample. Warm the tube, cool, and then add a few drops of 10% sodium hydroxide. A deep orange color confirms the presence of protein.
6. See Activity 72, “Detecting Nitrogen in Hair”, and Activity 83, “What Is a Cooked Egg?”

ANSWERS TO THE QUESTIONS

1. Milk, rice, fish, cooked egg, and gelatin.
2. Answers will vary.
3. Color intensity is proportional to concentration.
4. The protein is digested into amino acids, which can be restructured in your cells into your proteins.

76. Starch in Plant Parts

Plant leaves and other green parts use the light energy from the sun to produce simple sugars out of carbon dioxide in the air and water from the roots. Some of the sugars are linked together to form starch molecules, which are stored in the plant. Is starch present in plant leaves, roots, and stems? You will investigate the presence of starch in various parts of the plant.

Materials

1. Common plants including vegetables and fruits:
   A. Leaves: cabbage, lettuce, house plants, or leaves of different colors.
   B. Stems: white potatoes, broccoli stems, or twigs.
   C. Roots: sweet potatoes, carrots, or other roots.
2. Iodine solution.
3. Test tube.
4. Beaker, 100 mL or 250 mL.
5. Heat source.

Procedure

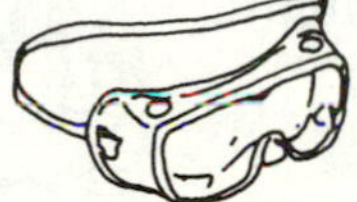

1. Arrange the plant samples in the three categories: leaves, stems, and roots.
2. Cut the part open to expose the interior so that the cell contents can be tested.
3. For leaves, you may need to remove the waxy cuticle: Boil the leaf for 5 min in a beaker full of water. Place 5 mL of ethyl alcohol in a test tube in the water and place the leaf in the alcohol. Allow the test tube to soak in the water bath for another 5 min. Pour out the alcohol and repeat the alcohol bath. Turn off all flames in the room while using the alcohol. Alcohol is flammable.
4. Try the test for starch by adding a drop of iodine solution to a drop of laundry starch or cornstarch. A blue-black complex is formed when starch and iodine come together.
5. Test each piece of plant (i.e., leaf, stem, and root) with a few drops of iodine solution.
6. Arrange the data to report which parts of plants contain the most stored starch.

Reaction

Starch is a storage molecule made of *glucose* units that are linked into coiled and branched chains. The iodine molecule is trapped in the branches and coils, and the color changes to blue-black and purple. Several thousands of these glucose units are in each starch molecule. Each glucose unit has the formula $C_6H_{12}O_6$.

Questions

1. Which plant part seemed to have the most starch?
2. Can you explain why starch would be concentrated in certain parts of a plant?
3. Design an experiment to test the change in starch concentration in leaves of a plant over time.

Notes for the Teacher

BACKGROUND

As a result of photosynthesis, glucose molecules, $C_6H_{12}O_6$, are produced. The glucose molecules are used by the plant for three main purposes: (1) to provide chemical energy in the cells; (2) to link up into cellulose molecules, which form the structure of the plant; and (3) to link up into starch molecules as storage of glucose for later use. The student can find the parts of the plant where most of the starch is stored by performing the iodine test. Most starch is stored in the roots; however, some is produced in the leaves and stored there. The amount found in the leaves depends upon the type of plant, the time of year, and the recent photosynthetic activity.

SOLUTIONS

The iodine solution is prepared by dissolving 2.5 g of potassium iodide in 500 mL of water and then adding 1.0 g of iodine crystals and stirring until they dissolve. Place in small dropper bottles. You may also dilute tincture of iodine (alcohol solution) from the drugstore with water.

TEACHING TIPS

1. You may want to also provide some fruits and grains (bread). See Activity 87 for the comparison of a ripe banana with an unripe one.
2. The leaf needs to be boiled to soften the waxy cuticle and the cell walls.
3. The alcohol will dissolve the wax on the leaf and the pigment. The pigment, when removed, cannot mask the positive starch test.
4. You might have students design tests by which they can find which variables result in different concentrations of starch. Leaves kept in the dark for 24 h will differ greatly from the leaves of a plant that has been brightly illuminated.
5. The blue color is produced when the I_5^- or I_3^- complex fits inside the coiled molecule of amylose, one of the two components of starch.

ANSWERS TO THE QUESTIONS

1. The roots of most plants.
2. The roots are the storage organ for plant food.
3. A student might collect a leaf at different times of the day or at different times of the year but the same time of day. The student should see the need to control all variables except one for each test.

Chemistry of Foods

77. What's in Baking Powder?

When you make biscuits or pancakes, you like them to be light and fluffy. Using baking powder in the recipe will result in a chemical reaction that makes gas bubbles that cause your batter to rise. Baking powder is a mixture of baking soda and an acid in solid form. When moist, the soda and acid react to form carbon dioxide gas. You will perform qualitative analysis to determine which of three common acids is in your sample of baking powder. The common sources of acids used in baking powders are a tartrate, a hydrogen phosphate, or an alum (sodium aluminum sulfate).

Materials

1. Fresh baking powder (sample 1, 2, or 3).
2. Filter paper and funnel.
3. Test tubes.
4. Heat source.
5. Barium chloride ($BaCl_2$) solution.
6. Hydrochloric acid (HCl), 0.1 M.
7. Nitric acid (HNO_3), 0.1 M.
8. Ammonium molybdate [$(NH_4)_2MoO_4$] solution.

Procedure

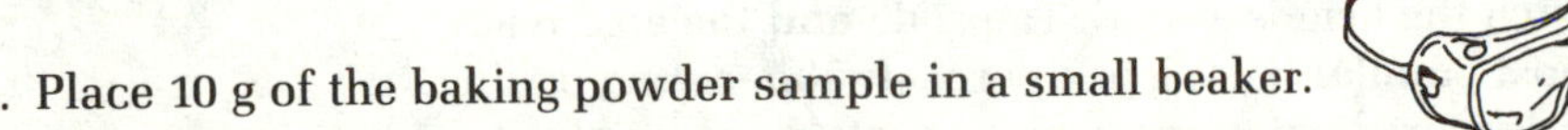

1. Place 10 g of the baking powder sample in a small beaker.
2. Add 50 mL of water.
3. Observe the gas. For a test for carbon dioxide gas, see Activity 6.
4. Set up a funnel and filter paper.
5. When the foaming stops, pour the solution through the filter paper.
6. Using the solution that has passed through the filter paper (the filtrate), prepare test tubes 1, 2, and 3 as follows:
   A. Place 5 mL of filtrate in each of three test tubes.
   B. To test tube 1, add 3–5 drops of $BaCl_2$ solution and 3–5 drops of HCl. If a white precipitate ($BaSO_4$) remains, then the baking powder acid was alum (sodium aluminum sulfate). Avoid contact with $BaCl_2$ solution. It is toxic.
   C. To test tube 2, add 3–5 drops of HNO_3 solution and 3–5 drops of $(NH_4)_2MoO_4$ solution. Warm in hot water. If a yellow precipitate [$(NH_4)_3PO_4 \cdot 12MoO_3$] forms, then the baking powder acid was calcium phosphate.
   D. Place the contents of test tube 3 in an evaporating dish and heat. If a black-charred residue (carbon) results, then the baking powder acid was potassium acid tartrate.

Reactions

Test Tube 1:

$$Ba^{2+}(aq) + SO_4^{2-}(aq) \longrightarrow \underset{\text{white}}{BaSO_4(s)}$$

Test Tube 2:

$$PO_4^{3-}(aq) + 12MoO_4^{2-}(aq) + 24H^+(aq) + 3NH_4^+ \longrightarrow \underset{\text{yellow}}{(NH_4)_3PO_4 \cdot 12MoO_3(s)} + 12H_2O$$

Test Tube 3: Because the products are a mixture of substances, an equation would be quite complex and not helpful here.
Potassium acid tartrate ($KHC_4H_4O_6$) decomposes when heated to form carbon (C) and water (H_2O) as well as other species. The black carbon identifies the molecule that is partly tartrate.

Questions

1. What gas is produced when the baking powder is moist?
2. On the basis of your results in this experiment, what kind of baking powder do you think you had?
3. Is baking powder a compound or a mixture?
4. Explain how you decided which acid your baking powder contained.

Notes for the Teacher

BACKGROUND

Most breads and cakes have small gas bubbles incorporated in the batter to cause the baked product to be leavened. Most often the gas is carbon dioxide produced by the metabolism of yeast or the chemical reaction between baking soda ($NaHCO_3$) and an acid. Often the acid is in buttermilk, fruit juice, or vinegar. Another source of acid is baking powder in which the acid is a solid. The soda and the acid react when they become wet. The most popular baking powder is double-acting baking powder. It contains a slow-acting ingredient, alum, which reacts more quickly when warm (in the oven). It also contains another acid, usually a hydrogen phosphate, which reacts more quickly in the mixing of the batter.

SOLUTIONS

1. The concentrations of the ammonium molybdate and the barium chloride solutions are not critical. Add 4 g of ammonium molybdate to 100 mL of water. Also add 4 g of barium chloride to 100 mL of water.
2. Hydrochloric acid and nitric acids are 0.1 M.
3. The dilutions of the acids are described in Appendix 2.

TEACHING TIPS

1. This activity serves as a good introduction to the ideas of chemical reactions, mixtures, compounds, formulas, and analysis.
2. Look at the ingredients in the baking powders on the market when deciding which samples to buy.
3. See Activity 81 for another test for the presence of a phosphate.

ANSWERS TO THE QUESTIONS

1. Acid or water reacts with sodium bicarbonate to produce carbon dioxide.
2. Answers will vary depending upon the type of baking powder.
3. Baking powder is a mixture.
4. By observing whether or not precipitates formed in test tubes 1, 2, and 3.

78. Sulfur Dioxide in Foods

Sulfur is burned to produce sulfur dioxide, which is used to bleach and preserve foods, especially fruits. It prevents the growth of bacteria and mold. Apples, pears, apricots, peaches, and figs are dried and preserved with SO_2. Although packagers try to remove as much of the sulfur compounds as possible, some usually remains. You will test for the presence of residual sulfur compounds by reacting fruit with barium chloride to form a new compound that is not soluble. A substance that is not soluble will form a cloudy precipitate. You will compare fruit preserved with sulfur and without sulfur. **This activity will take 2 days.**

Materials

1. Sulfured dried fruit from grocery store.
2. Unsulfured dried fruit from health food store.
3. Hydrogen peroxide (H_2O_2), 3%.
4. Barium chloride ($BaCl_2$) solution, 3 M.
5. Funnel with stand.
6. Four beakers, 250 mL.
7. Stirring rod.

Procedure

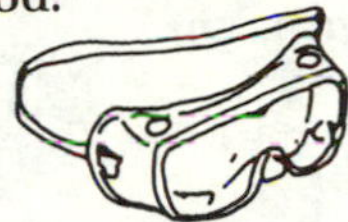

1. Overnight, soak several slices of each kind of dried fruit in distilled water. Use enough water to cover the slices.
2. Place several thicknesses of cheesecloth in a funnel, or use a coffee filter. Pour the liquid through the funnel, and then place the fruit in the funnel to squeeze out as much liquid as possible. Use a stirring rod. Do this step for each of the two samples.
3. Add 25 mL of 3% H_2O_2 to each filtrate in the beakers under the funnel to oxidize the sulfur atoms so that they can be identified as described in the next step.

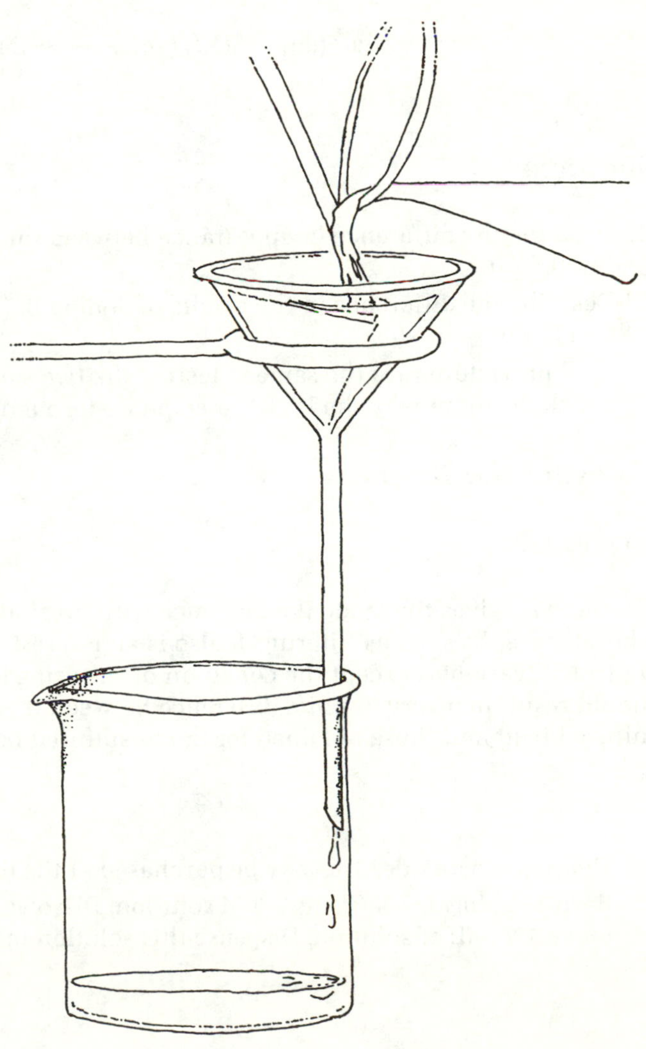

4. To separate the sulfur (in the form of sulfate) from the solution, add $BaCl_2$ drop by drop. Be careful with $BaCl_2$. It is toxic. Avoid contact with the solution.
5. You may want to filter the barium sulfate precipitate, dry it, and weigh it. If you had weighed the fruit at the beginning, you could determine the percentage of sulfur in the fruit.

Reactions

When sulfur dioxide dissolves in the moisture in the fruit, it reacts to form the sulfite ion (SO_3^{2-}):

$$SO_2(g) + H_2O(\ell) \longrightarrow SO_3^{2-}(aq) + 2H^+(aq)$$

The sulfite ion reacts with hydrogen peroxide (H_2O_2) to form the sulfate ion (SO_4^{2-}):

$$\underset{\text{sulfite}}{SO_3^{2-}(aq)} + H_2O_2(aq) \longrightarrow \underset{\text{sulfate}}{SO_4^{2-}(aq)} + H_2O(\ell)$$

The sulfate ion reacts with the barium ion in the $BaCl_2$ to produce the precipitate, barium sulfate ($BaSO_4$):

$$Ba^{2+}(aq) + SO_4^{2-}(aq) \longrightarrow \underset{\text{white}}{BaSO_4(s)}$$

Questions

1. Describe the difference in appearance between the sulfured fruit and the unsulfured fruit.
2. Describe the difference in the results of adding $BaCl_2$ to the solutions from the two fruits.
3. If all procedures are the same in testing the two samples, what can you conclude about why the $BaSO_4$ precipitate formed?

Notes for the Teacher

BACKGROUND

This activity gives the student experience with precipitation, solubility, and some laboratory skills such as filtering. It also is an interesting experience in reasoning: to control all variables except the condition of sulfuring in the fruit. This condition should result in a very definite difference between the results obtained for the sulfured fruit and those obtained for the unsulfured fruit.

SOLUTIONS

1. Hydrogen peroxide, 3%, may be purchased at the drugstore.
2. Barium chloride ($BaCl_2$) is a 3 M solution. Dissolve 62 g of $BaCl_2$ in water to make 100 mL of solution. Dispense this solution in dropper bottles.

TEACHING TIPS

1. Barium compounds are toxic. Keep this solution in small dropper bottles.
2. You might soak the fruits in two large beakers for the class. Each student could take a couple of pieces and some juice to filter.
3. Hydrogen peroxide (H_2O_2) is an oxidizing agent. The sulfur atoms are *oxidized*, which means that electrons are lost. The sulfur atom in sulfate (SO_4^{2-}), which has lost six electrons, has fewer electrons than the sulfur atom in sulfite (SO_3^{2-}), which has lost four electrons. Sulfur atoms that have lost no electrons form a yellow solid.
4. Students should see no precipitate of $BaSO_4$ in the unsulfured solution and a fine white precipitate in the sulfured solution.
5. Check the label on the dried-fruit package to be sure that it contains sulfur dioxide (SO_2). Because 13.7% of $BaSO_4$ is sulfur, measuring the mass of the fruit before the experiment and the precipitate after the experiment will allow the students to find the amount of sulfur in the fruit and the percent of sulfur in the fruit. These determinations might be suggested for capable students.
6. Do not allow students to eat any of the food in this activity.

ANSWERS TO THE QUESTIONS

1. Unsulfured fruit is brown. Sulfured fruit is white.
2. A white precipitate forms in the sulfured solution.
3. The sulfured fruit contains sulfur.

79. Calcium and the Coagulation of Milk

One way to coagulate milk for making cheese is to add an enzyme called *rennin*. Calcium ions must be present for this enzyme to work. You will investigate the role of calcium in the coagulation of milk by adding rennin. You will design an experiment to test for the relationship between calcium ions and coagulation.

Materials

1. Fresh milk.
2. Rennin (e.g., Rennet).
3. Sodium citrate, solid.
4. Calcium chloride ($CaCl_2$) solution, 1%.
5. Three test tubes and stoppers.
6. Droppers.
7. Water bath.
8. Thermometer.

Procedure

1. Prepare a warm water bath at 37 °C.
2. Fill test tubes 1, 2, and 3 half full with fresh milk.
3. Add a pinch of rennin to tube 1.
4. Add a pinch of rennin and a pinch of sodium citrate to tube 2.
5. Tube 3 is a control. Add nothing to the milk in it.
6. Place the tubes in the water bath.

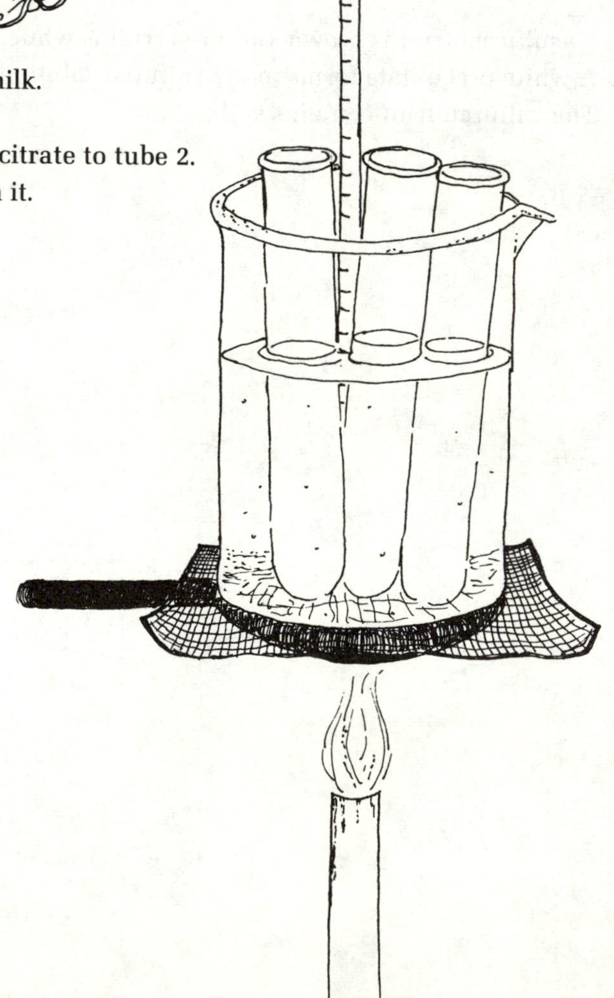

7. After 5 min, examine the tubes.
8. Using the 1% calcium chloride solution, how could you prove that calcium ions are needed for the activation of rennin?
9. Test the hypothesis with your suggested procedure.

Reaction

Rennin converts the milk protein caseinogen into paracasein. Calcium reacts with the paracasein to form "unbalanced" molecules, and the milk coagulates.

Calcium ions can be removed from milk by the addition of sodium citrate, which reacts with the calcium ions:

$$Na_3C_6H_5O_7(aq) + Ca^{2+} \longrightarrow NaCaC_6H_5O_7(s) + 2Na^+$$

When the calcium ions are not available, the paracasein molecules cannot cause coagulation.

Questions

1. Explain the coagulation process.
2. How did you test for the role of calcium ions in coagulation?
3. What happens to the calcium ions when sodium citrate is added to milk?

Notes for the Teacher

BACKGROUND

Milk is considered high in protein, although only 3.5% by weight is protein. Most of the protein is in the form of casein molecules that normally clump together into small spherical structures called *micelles*. In cheese manufacture, the rennin prepares the micelle for calcium ions to enter and disrupt the normal casein structure. If calcium is removed, this process cannot take place, and the milk will not coagulate. This activity allows the student to discover this relationship and to design a simple way to test the role of calcium in milk coagulation.

SOLUTION

Calcium chloride ($CaCl_2$) is approximately 1%. Dissolve 1 g in 100 mL of water.

TEACHING TIPS

1. A large beaker of warm tap water will remain at 37 °C for 5 min and may be used by several teams. If you begin with the temperature around 45 °C, adding the tubes will cool the water to the desired temperature.
2. Rennin is readily available in the Junket product called Rennet. It contains added calcium salts; thus, if you use it, your students may need to add a large pinch of sodium citrate during Procedure 4. You can also use Rennilase, purchased from any biological supply company.
3. This activity is a good way to emphasize the need to control variables and to include a control trial in the samples.
4. This activity is also a good way to emphasize the chemical nature of all substances and common materials.

5. Students should have little trouble deciding that to test the role of calcium (Procedure 8), they should add calcium ions to tube 2 to observe whether the presence of calcium results in a change.
6. Adding calcium ions to tube 2 after the sodium citrate has been added should result in coagulation. The ideal number of drops is 5–10, depending on the source of rennin you use. You might have each team use a different number of drops to test the optimum number for your conditions. Be sure to control the temperature.
7. Do not allow students to eat or drink any of the food in this activity.

ANSWERS TO THE QUESTIONS

1. See the Reaction section.
2. By adding calcium ions to tube 2.
3. The calcium ion is tied up in the molecule, sodium calcium citrate.

80. Vitamin C in Foods

Vitamin C is also called *ascorbic acid* and is an important nutrient in the human diet. It is part of enzyme systems that control chemical reactions in your cells. The kind of reaction that uses ascorbic acid is called an *oxidation–reduction* reaction or *electron-transfer* reaction. Vitamin C is found naturally in many foods, especially citrus fruits. You will look for the relative amounts of vitamin C in various juices by allowing the ascorbic acid to react with iodine. When all the ascorbic acid is reacted with iodine, the iodine added beyond that point will make a purple complex with starch.

Materials

1. Fruit juices: fresh, frozen, powdered, and canned.
2. Vitamin C tablet, 100 mg (0.100 g).
3. Starch solution, 1%.
4. Iodine solution.
5. Hydrochloric acid (HCl), 1 M.
6. Buret.
7. Graduated cylinders, 50 and 100 mL.
8. Erlenmeyer flask, 250 mL.

Procedure

1. Make a standard solution of ascorbic acid by dissolving one vitamin C tablet in 100 mL of water.
2. Place 30 mL of the vitamin C solution in a flask.
3. Add 2 drops of HCl. Add 2 droppers full of starch solution.
4. Fill the buret with the iodine solution. Record the initial volume reading.
5. Slowly add the iodine to the flask while swirling the flask until the solution stays blue-black for 15 s.

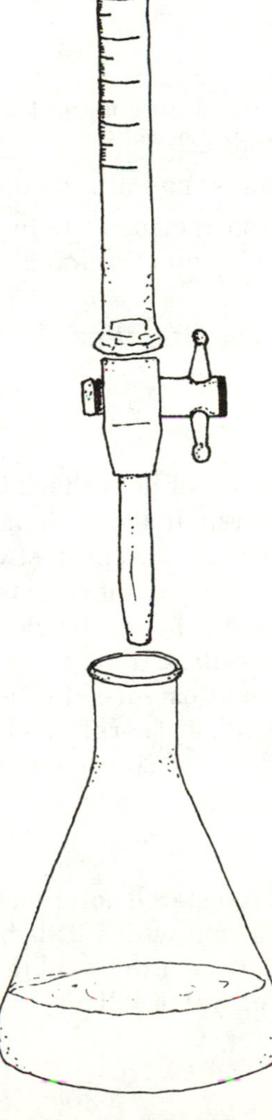

6. Record the volume reading on the buret. If you began at 0.00, the volume you read on the buret will be the volume of iodine solution you have added to the flask.
7. Using uniform amounts of juices, add iodine solution from the buret until all the ascorbic acid is reacted in each juice (when the solution stays blue-black for 15 s). Remember to record the initial volume of the buret each time.
8. Compare the relative amounts of ascorbic acid present in the samples you are testing. Compare your results with those of other members of the class. The more iodine that is required, the greater the concentration of vitamin C.

Reactions

When ascorbic acid molecules react with iodine molecules, the ascorbic acid is *oxidized* (loses electrons) and the iodine is *reduced* (gains electrons). Reduced iodine cannot react with starch. When all the ascorbic acid is gone, added iodine molecules are then able to react with starch. Therefore, the amount of iodine added before the blue-black is seen is directly related to the amount of ascorbic acid in the solution.

$$I_2 + \text{ascorbic acid} \longrightarrow \text{oxidized ascorbic acid} + 2I^-$$

$$I^- + I_2 + \text{starch} \longrightarrow I_3^-\text{-starch complex (blue-black)}$$

Oxidized ascorbic acid in fruit juice is produced when the juice is in contact with oxygen in the air. Oxidized ascorbic acid is not useful to the body as a vitamin.

Questions

1. How do you know when the ascorbic acid is gone from the solution as you are adding iodine?
2. Which juices contain the most vitamin C? Which contain the least?
3. Design an experiment to test the effect of air, heat, or other substances on the vitamin C content of foods.

Notes for the Teacher

BACKGROUND

Vitamin C is required in the daily diet because it is a water-soluble substance and cannot be stored. It also cannot be synthesized by human cells. It is involved in tissue repair reactions, protective oxidation reactions in cells, and reactions affecting the integrity of cell membranes and connective tissue. As vitamin C readily reduces other molecules, it becomes oxidized. This function is the role it must play in cells. If a molecule of ascorbic acid has been oxidized by oxygen in the air or another substance, it is no longer available to a cell. In this activity students purposely oxidize ascorbic acid molecules with iodine in order to find out how much ascorbic acid is in a sample.

SOLUTIONS

1. Prepare the starch solution by mixing ½ tsp of soluble starch (cornstarch or potato starch) with a little bit of water. Stir this mixture into 100 mL of boiling water. Boil for 1 min and then cool. One teaspoon of liquid laundry starch in 100 mL of water will also work.

2. Prepare the iodine solution by dissolving 0.6 g of potassium iodide (KI) in 500 mL of water. Then dissolve 0.6 g of iodine in 50 mL of ethyl alcohol. Mix the two solutions and dilute to 1.0 L with water.
3. See Appendix 2 for the preparation of hydrochloric acid.

TEACHING TIPS

1. Because you are looking for a color change, avoid dark juices such as grape juice.
2. Fruit juices should be opened or prepared right before use. You may want to introduce the variable of exposure time by allowing air to come into contact with a sample overnight or by boiling a sample and then cooling it.
3. A class of students should be able to gather data about the relative vitamin C content of many foods and juices. You might include the variables of air exposure and heat so that each team is investigating a different variable or a different juice. You can then rank the findings.
4. Minimum daily requirement (U.S. RDA) for vitamin C is 0.060 g/day. Linus Pauling, winner of two Nobel prizes, has become a controversial figure because of his recommendation for a much higher intake of vitamin C.
5. If you do not have burets, you can make the iodine solution more concentrated and add the iodine by counting the drops.
6. The formula for ascorbic acid is $C_6H_8O_6$.
7. The blue color for starch is formed when the I_5^- or I_3^- complex fits inside the coiled structure of amylose, one of the two components of starch.
8. Do not allow students to drink any of the liquids in the chemistry laboratory.

ANSWERS TO THE QUESTIONS

1. The added iodine suddenly reacts with the starch to make it blue-black.
2. Answers will depend on conditions.
3. Experiments should be similar to this one.

81. Detecting Phosphates in Cola

One of the main ingredients in many cola beverages is phosphoric acid. This compound helps to preserve the beverage, gives it part of its sweet taste, and provides an acidic solution. Phosphoric acid *ionizes* to hydrogen ion, H^+, and phosphate ion, PO_4^{3-}, in cola. In this activity we will test for the presence of phosphates in cola and also in common materials such as laundry detergent.

Materials

1. Cola soft drink.
2. Laundry detergent.
3. Sodium phosphate (Na_3PO_4) (or other phosphate salt).
4. Magnesia mix.
5. Test tubes.
6. pH paper.
7. Hydrochloric acid, 1.0 M.
8. Droppers.

Procedure

1. Establish a positive test for phosphate using sodium phosphate.
   A. Place 0.5 g of sodium phosphate (Na_3PO_4) in a test tube. Fill the tube half full with distilled water and shake the tube until the solid dissolves.
   B. Wet a piece of pH paper (or red litmus paper) with the solution. It should be basic.
   C. Add hydrochloric acid, 1 drop at a time, until the solution is neutral.
   D. Add a dropper full of magnesia mix to the tube.
   E. Set the tube aside and observe it carefully. A white precipitate that slowly forms and settles to the bottom of the tube confirms the presence of phosphate.
2. Test cola for phosphate.
   A. Repeat the previous test, but use a test tube half-filled with cola instead of distilled water.
   B. The cola may already be acidic, so it may not be necessary to add hydrochloric acid.
3. Test detergent for phosphate.
   A. Repeat the previous test, but dissolve a small amount of granular detergent (enough to cover a dime) in a test tube half-filled with water.
   B. Be sure to test the solution with pH paper. It may be necessary to add acid to make the solution neutral.

Reaction

Magnesia mixture consists of magnesium chloride ($MgCl_2$), ammonium chloride (NH_4Cl), and aqueous ammonia (NH_4OH). When this mixture is added to an acidic solution containing a phosphate ion, a white precipitate, magnesium ammonium phosphate ($MgNH_4PO_4$) is formed. This result confirms the presence of phosphate:

$$Mg^{2+}(aq) + NH_4^+(aq) + HPO_4^{2-}(aq) \longrightarrow \underset{\text{white}}{MgNH_4PO_4(s)} + H^+(s)$$

Questions

1. How did you show the presence of phosphate ion in the samples?
2. Show the formula for phosphate.
3. How did the amount of phosphate in the cola and the laundry detergent compare?
4. Which mineral is provided by the phosphate ion?

Notes for the Teacher

BACKGROUND

Phosphorus is one of the major minerals in the body. It is necessary for proper bone development, and a deficiency can cause serious defects. One of the primary sources of phosphorus is phosphates, and phosphates are in high concentration in cola beverages because of the phosphoric acid added during the bottling process. In this activity students will have an opportunity to first establish a qualitative test for the presence of phosphates and then use this test to show the presence of phosphates in cola and also in some laundry detergents.

SOLUTIONS AND MATERIALS

1. Hydrochloric acid (HCl) is 1.0 M. See Appendix 2.
2. Sodium phosphate (Na_3PO_4): This compound is used to establish a test for a known phosphate. You can either use the solid or prepare a 10% solution.
3. Magnesia mix: This solution is used exclusively to test for phosphates.
   A. Add 10.0 g of magnesium chloride ($MgCl_2$) to 100 mL distilled water and stir until all of it dissolves.
   B. Dissolve 25.0 g of ammonium chloride (NH_4Cl) in the magnesium chloride solution.
   C. Add 100 mL of 6 M ammonium hydroxide (aqueous ammonia, NH_3) (see Appendix 2 for diluting NH_3), dilute the total solution to 350 mL with distilled water, and let the solution stand for 2 h before using it.
      Because only a few milliliters of this solution will be needed by each group of students, it can be conveniently dispensed in small dropper bottles.

TEACHING TIPS

1. The phosphate ion is PO_4^{3-}. Hydrogen phosphate is HPO_4^{2-}, and dihydrogen phosphate is $H_2PO_4^{-}$.
2. Monosodium phosphate (NaH_2PO_4) is used in many foods to control acidity.
3. Calcium dihydrogen phosphate [$Ca(H_2PO_4)_2$] is the acid-producing substance in some baking powders. This compound is also used in fertilizers to increase the amount of phosphate ion, which is essential to plant growth.
4. Sodium triphosphate ($Na_5P_3O_{10}$) is the primary phosphate in detergents. This compound acts to break up and suspend dirt particles by forming complexes with metal ions.
5. In the late 1960s, a serious problem occurred with high-phosphate detergents. These entered lakes and streams and produced an overgrowth of algae and other organisms. This process is called *eutrophication*. Laws enacted in the 1970s reduced the levels of phosphates in detergents.
   The acceptable phosphate level in detergents is now set by law as 7.1% phosphorus as phosphates. This level is about 6.0 g per cup of detergent.

6. If your students have difficulty seeing the white precipitate in the test tubes, you might try diluting the cola or try to remove the color with activated charcoal (see Activity 94).
7. You might have students do a comparative investigation of several brands of soft drink. Some brands do not contain phosphoric acid. Do not allow the students to drink anything in the chemistry laboratory.

ANSWERS TO THE QUESTIONS

1. By forming a precipitate with the phosphate ion, ammonium ion, and magnesium ion ($MgNH_4PO_4$).
2. Phosphate is an ion (PO_4^{3-}).
3. The answer depends upon the detergent used. More precipitate will probably be formed with cola.
4. Phosphorus.

82. Artificial Flavorings and Fragrances

Many foods and candies contain artificial flavorings. Many of these flavorings are molecules called *esters* that are synthesized in the laboratory. These fruity and delicious flavors for foods are made by reacting alcohols with acids. You will prepare several kinds of fragrant ester molecules. Naming esters is fun, and you will develop this skill also.

Materials

1. Salicylic acid ($C_7H_6O_3$).
2. Butyric acid ($C_4H_8O_2$).
3. Decanoic acid ($C_{10}H_{20}O_2$).
4. *n*-Amyl alcohol (1-pentanol, $C_5H_{11}OH$).
5. *n*-Octyl alcohol (1-octanol, $C_8H_{17}OH$).
6. Methyl alcohol (methanol, CH_3OH).
7. Ethyl alcohol (ethanol, C_2H_5OH).
8. Beaker for water bath.
9. Test tubes.
10. Graduated cylinder, 10 mL.
11. Hot plate.
12. Aluminum foil.
13. Glacial acetic acid (100% acetic acid, CH_3COOH).
14. Sulfuric acid (H_2SO_4), concentrated.

Procedure

CAUTION: Be careful with the acetic and sulfuric acids. Wear gloves, goggles, and a face shield. Also extinguish all flames in the room because the alcohols are flammable.

1. Prepare a water bath by filling a beaker half full with water and heating it.
2. Look at the table to find an acid and an alcohol that you will combine.

Tube	*Alcohol*	*Acid*	*Ester Fragrance*
1	*n*-amyl alcohol	butyric acid	apricot
2	*n*-amyl alcohol	salicylic acid	pineapple
3	*n*-amyl alcohol	glacial acetic acid	banana
4	*n*-octyl alcohol	glacial acetic acid	fruity
5	methyl alcohol	salicylic acid	wintergreen
6	ethyl alcohol	decanoic acid	grape
7	ethyl alcohol	butyric acid	apple
8	ethyl alcohol	glacial acetic acid	fruity

3. Place 2 mL of the alcohol selected from the table for each ester chosen in a test tube. Label the tube.
4. If the acid is a liquid, add 2 mL of it to the alcohol in the tube. If it is a solid, add 1 g of it.
5. Carefully add 1 mL of concentrated sulfuric acid to the tube. Your teacher may choose to pour this acid for you. Wear gloves, goggles, and a face shield.
6. Tap the bottom of the tube with your finger to mix.
7. Place tubes 1 and 2 in the water bath.

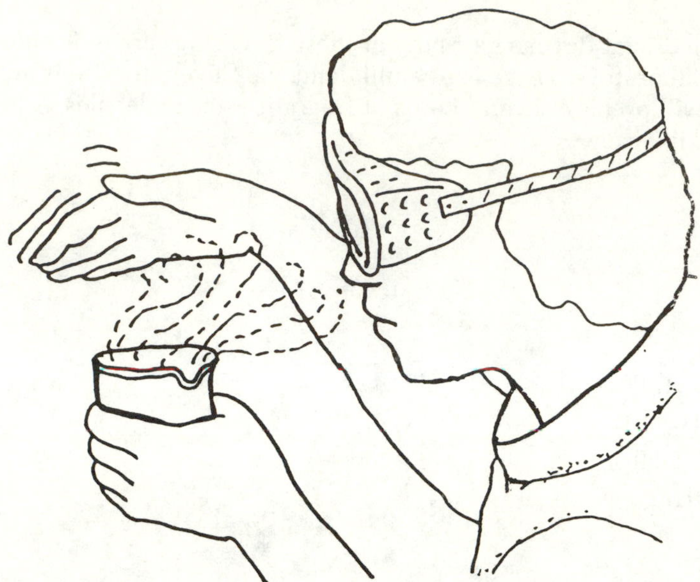

8. After a few minutes, check the tubes for a fragrance. If no fragrance is present, cover the tube with aluminum foil, set it aside, and check again the next day.

Reaction

An ester molecule is formed when an alcohol molecule and an organic acid molecule come together and a water molecule is released:

$$\underset{\text{organic acid}}{RCOOH} + \underset{\text{alcohol}}{HOR'} \longrightarrow \underset{\text{ester}}{RCOOR'} + \underset{\text{water}}{H_2O}$$

The carbon group R stands for the rest of the acid molecule, and the carbon group R′ stands for the rest of the alcohol molecule.

Esters are named by first giving the alcohol name and then the acid name, which is changed to have an -ATE suffix instead of an -IC suffix. For example, ethyl alcohol with acetic acid produces ethyl acetate.

Read the labels on artificially flavored foods and candies to find additives that are esters.

Questions

1. Describe the odor of the ester produced in each tube you tested.
2. Use the guide in the Reaction section to write the reaction for the ester you produced.
3. Name the ester you produced.

Notes for the Teacher

BACKGROUND

Many esters are fragrant oily molecules. Investigators discovered about 200 years ago that such fragrances could be produced when an alcohol was reacted with an organic acid. The definite fragrance or aroma of most common foods is due to some combination of natural esters. The aroma of freshly ground coffee is due to more than 200 compounds, many of them esters. Perfumes are esters. However, many other common esters do not have a fragrance, for example, aspirin, fats, and oils.

TEACHING TIPS

1. Concentrated acids, namely, glacial acetic acid and sulfuric acid, are used in this procedure. The teacher should dispense these acids and keep them in the hood for this purpose. Because of the small volume of each used, it is wise to estimate the volume rather than to use a graduated cylinder. Dispense from a small beaker or from a separatory funnel or buret suspended over a dish of sand.
2. The sulfuric acid serves as a *dehydrating agent.* If water molecules are removed from the products, the reaction to produce the ester favors the making of more ester. This situation is a principle of equilibrium because the reaction is reversible.
3. Caution students to agitate the tubes gently so that the acids do not spill out.
4. Students should use a hand to waft fumes toward them as they smell the reactions. Demonstrate this technique for them. They should notice, also, the odors of the reactants before the reactions.
5. Some of these reactions take place quickly. Therefore, students should check quickly for odors.
6. If odors are not detected, cover the tubes with aluminum foil until the next day.
7. Immediate results are usually found with methyl salicylate (wintergreen) and ethyl acetate (fruity).
8. Butyric acid produces the odor of rancid butter. You may wish to use it so that the students can observe the dramatic difference in odor between the reactants and products. However, you may not want very many teams to choose this one.
9. The organic acids and alcohols should not be discarded in the sink. Some are not very water-soluble. These small volumes may be poured onto paper towels and dried in the hood. Avoid flames.

ANSWERS TO THE QUESTIONS

1. Amyl butyrate, apricot; amyl salicylate, pineapple; amyl acetate, banana; octyl acetate, fruity; methyl salicylate, wintergreen; ethyl decanoate, grape; ethyl butyrate, apple; and ethyl acetate, fruity (apple).
2. If you have students answer this question, they will need to consult textbooks to find the structures of the acids and alcohols.
3. See answer 1.

83. What Is a Cooked Egg?

Everyone is familiar with the change that occurs when an egg is cooked. It solidifies. But why? Strangely enough, the chemistry involved in the change that *coagulates* the egg protein is an unsolved mystery. This process apparently involves the unraveling of long protein molecules. As they unravel, they can link up with each other in new ways and form the solid. Several agents cause the coagulation of protein. You will study some of them in this activity.

Materials

1. One fresh egg white (albumin).
2. Strong tea or 10% tannic acid solution.
3. Lead nitrate [$Pb(NO_3)_2$] solution, 10%.
4. Ethyl alcohol (95%) or isopropyl (rubbing) alcohol.
5. Droppers.
6. Test tubes, medium to large.
7. Beaker, 200 mL.
8. Hot plate.
9. Thermometer.

Procedure

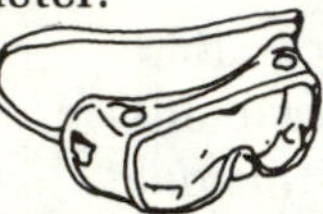

1. Stir the white of one egg into 100 mL of water.
2. Share with another team.
3. Fill five test tubes half full with the egg white solution.
4. Treat the test tubes as follows, and make observations and comparisons:
   Tube 1: Add nothing.
   Tube 2: Heat the tube for several minutes in a hot water bath on a hot plate, beginning at 50 °C. Record the temperature.
   Tube 3: Add 10 mL of alcohol. Alcohol is flammable. Extinguish all flames in the room.
   Tube 4: Add 10 mL of strong tea or tannic acid solution.
   Tube 5: Add several drops of 10% $Pb(NO_3)_2$ solution. Caution: Avoid contact with lead nitrate. It is toxic.

Reaction

Each substance coagulated the protein. The structure of the protein was changed, and a new molecular structure was formed. We say that the protein was *denatured.*

Questions

1. How do the egg albumin samples in each of the tubes compare with the samples in tubes 1 and 2?
2. At what temperature did the egg albumin coagulate in tube 2?
3. Why do you think the skin is swabbed with alcohol before an injection?
4. What would be the usefulness of tannic acid in the tanning process of hides?
5. Why would egg white be a treatment for people who have ingested toxic lead salts?

Notes for the Teacher

BACKGROUND

Protein molecules are large structures with up to thousands of amino acid residues in specific sequences. Twenty kinds of amino acid are common, although more than 150 different amino acids exist. The first sequence to be found for a particular protein was the sequence of 51 amino acids in sheep insulin; Frederick Sanger found this sequence in the 1950s. Since then, scientists have found the sequences of amino acids in many proteins. Catalase, an enzyme protein, has a sequence of 2202 amino acids, for instance. However, the sequence is only part of the picture. As illustrated by this activity, protein chains are formed into very specific shapes that can be changed (denatured) under certain conditions. When certain substances, or heat, affect the protein, various bonds causing the specific shape are broken, and the protein chains tangle together in a disorganized aggregate. We say the denatured protein has *coagulated.* The primary attachments that are broken are the hydrogen bonds. In human nutrition, proteins must be denatured and broken down into amino acids. This process is accomplished by enzymes.

SOLUTIONS

1. Tannic acid is about 10%. Dissolve 10 g of tannic acid in 90–100 mL of water. Any other acid can be used.
2. Lead nitrate [$Pb(NO_3)_2$] is about 10%. Dissolve 10 g of $Pb(NO_3)_2$ in 90–100 mL of water.
3. Isopropyl rubbing alcohol from the drugstore is usually a 70% solution in water. Other alcohols can be used.

TEACHING TIPS

SAFETY NOTE: Use care with the solutions. Wear plastic gloves. All are toxic in this activity. Collect all of the samples from test tube No. 5 and dispose of them as directed in Appendix 4B.

1. Powdered albumin is also available. Egg whites are almost pure protein (albumin).
2. A temperature of 50 °C will cause denaturation and precipitation of most proteins.
3. pH changes upset the slight positive and negative regions of the molecules and thus cause bonds to be broken and the specific shape to be destroyed. Positive regions form on the amino group and negative regions form on the acid group of amino acids, which are arranged as close to each other as the strands fold.
4. Tannic acid is used to coagulate the protein in hides to make them leathery and suitable for clothing. Tannic acid, found in strong tea, can also coagulate the protein on a burn site to form a protective coating to prevent infection.
5. Alcohol coagulates protein by interfering with the hydrogen bonds that contribute to the specific shape of the protein. A 70% solution in water will easily penetrate bacterial cells and denature and coagulate the protein of the bacterial cell. Rubbing alcohol is a 70% solution.
6. Heavy metal salts such as silver and mercury have been used as antiseptics in low concentration. Heavy metal ions such as Hg^{2+} and Pb^{2+} attach to sulfur groups on certain amino acids and thereby denature the protein. Because these "heavy metal" ions react with body proteins, including enzymes, they must be used with great care and not released into the environment.

7. You might have students vary the temperatures and concentrations of the agents under study in this activity to determine the thresholds of temperature, pH, and heavy metal concentration that will result in denaturation.

ANSWERS TO THE QUESTIONS

1. Tube 1 will be clear. The others will be white and cloudy.
2. Around 50–70 °C.
3. To coagulate bacterial protein.
4. To coagulate protein in animal skins.
5. Lead can coagulate the albumin instead of poisoning the victim. The albumin containing the lead must then be removed from the stomach.

84. Physical and Chemical Properties of Fresh Milk

We think of milk as an ordinary substance and may overlook its actual nature. Milk has been called "the perfect food" and is a complex mixture of substances in solution and fat globules in suspension. You will search for and identify the presence of several of these components by using physical means and standard chemical tests.

Materials

1. Fresh whole milk.
2. Vinegar (acetic acid, 5%).
3. Universal pH paper or litmus paper.
4. Sudan red dye.
5. Sodium hydroxide (NaOH) solution, 6 M.
6. Copper sulfate ($CuSO_4$) solution, 0.1%.
7. Heat source.
8. Test tubes.
9. Droppers.
10. Microscope and slides.

Procedure

1. Obtain a test tube of fresh milk.
2. Use universal pH paper or litmus paper to identify whether milk is acidic, neutral, or basic.
3. Place 1 drop of milk on a microscope slide and add 5 drops of water and 1 drop of Sudan red dye. Mix and examine under the microscope. Look for the red-dyed fat globules.
4. Add several milliliters of vinegar to the milk in the test tube, allow the curds to settle, and observe the coagulated protein (curds). Take note of the relative amount of the milk that is protein. If you wish, you may follow the procedure in Activity 50 to turn this product into casein glue.
5. Test for protein in milk by using the following procedure (a blue-violet solution indicates protein):
   A. Place 2 mL of fresh milk in a test tube.
   B. Add 2 mL of 6 M NaOH. Be careful: Wear gloves, goggles, and a face shield. Sodium hydroxide is a strong base.
   C. Add 1–10 drops of 0.1% $CuSO_4$ solution.

6. Determine the mineral content of milk by placing a known mass in a crucible; heat slowly and then strongly to form a white ash. The white ash is the mixture of minerals. The percentage of milk that is minerals can be determined after the mass of the mineral ash is found.

Questions

1. What is the pH (acidity) of fresh milk? What is the pH of sour milk?
2. What food products are based on the souring of milk?
3. Why do you think the red-dye fat globules "bounce" around?
4. What were two ways that you identified protein in milk?
5. What minerals might be in the ash you produced in Procedure 6?

Notes for the Teacher

BACKGROUND

Milk is analyzed at the dairy by a chemist and is generally found to be 87% water, 4.8% carbohydrate, 4% fat, 3.5% protein, and 0.7% minerals. The carbohydrate content is mostly the disaccharide sugar, lactose. Lactose acts as the energy source for bacteria that produce yogurt and buttermilk. The fat is in the form of globules that will be apparent under the microscope. The fat globules reflect light and give milk some of its whiteness. Some vitamins are dissolved in the fat. The protein is a mixture of proteins that are *complete*, which means that they have all essential amino acids. The most abundant protein is casein. This activity introduces students to the chemical composition and physical properties of milk. The student will see the role of the food chemist behind the scenes of the production of a common substance, milk.

SOLUTIONS

1. Sodium hydroxide (NaOH) solution is 6 M. Dissolve 24 g of NaOH in water to make 100 mL of solution.
2. Copper sulfate solution is about 0.1%. Dissolve 0.1 g of $CuSO_4 \cdot 5H_2O$ in 100 mL of water.
3. Sudan red dye is a fat-soluble dye. Add 1 g of this dye to 100 mL of ethyl alcohol.

TEACHING TIPS

SAFETY NOTE: Sodium hydroxide in the 6 M concentration is corrosive. Have students wear gloves and add the sodium hydroxide at a special station equipped with a face shield.

1. You may want to allow milk to sit at room temperature for several days before this activity so that some will be curdled by bacterial acids and can be compared with fresh milk.
2. You may want to test buttermilk and yogurt as well as milk.
3. Fresh milk is usually slightly acidic and has a pH of about 6.5 (pH 7 is neutral). If fresh milk is heated, the acid and heat cause the milk protein to coagulate at a rapid rate.
4. Students may enjoy imagining the cause of the Brownian motion seen against the red-dyed fat globules when the globules are viewed under the microscope. This activity is a chance to discuss the particulate nature of matter and the sizes of dissolved particles. The fat globules are jostled about by the dissolved substances in the milk.
5. Some students may want to research the milk sugar, lactose, to find how it is similar and different from table sugar, sucrose. Infants are born with the enzyme needed to utilize lactose, but they are several months old before they develop the enzyme to digest starch.
6. When vinegar is added to the milk, increased acidity causes the protein, casein, to clump. The molecules no longer repel each other; instead, they are attracted to each other in the more acidic medium. The large particles settle out as curds.
7. The test for protein is called the biuret test. See Activity 75, "Tests for Protein".

8. The mineral content of milk is usually less than 1% and may be difficult to detect with school balances. However, the appearance of the ash is obvious and valuable to observe.
9. Do not allow the students to consume any of the food in this activity.

ANSWERS TO THE QUESTIONS

1. pH 6.5 for fresh milk and 4.5 for sour milk.
2. Yogurt, buttermilk, sour cream, and cheese.
3. The movement occurs because of Brownian motion: the bombardment of large molecules by fast-moving, smaller molecules.
4. Protein was indicated by the test with acid to coagulate the protein and by the test with sodium hydroxide and copper sulfate, which is called the biuret test.
5. The minerals are in the form of inorganic salts; calcium phosphate is the most abundant.

85. How Much Fat Is in Ground Beef?

In this experiment we will determine the approximate amount of fat in ground beef. Grocery stores and meat markets usually mark the fat content on their packages of ground beef. The law requires that they provide this information. The law also permits no more than 30% fat in ground beef and 15% fat in lean ground beef.

Materials

1. Several samples of ground beef: ground beef (hamburger), ground lean beef, and ground chuck.
2. Balances.
3. Beakers, 600 mL.
4. Graduated cylinders, 100 mL.
5. Burners.

Procedure

1. Weigh a 100-g sample of your ground beef on a balance.
2. Place the beef sample in a 600-mL beaker.
3. Add 400 mL of water and stir it into the beef sample.
4. Place the beaker on a ring stand and carefully heat it until the water boils gently.
5. Continue to boil the meat for about 10 min.
6. Remove the beaker from the heat and allow it to cool. What happens to the meat? What happens to the fat?
7. Carefully pour off the top fat layer into a clean 100-mL graduated cylinder. If necessary, remove the last traces of fat with a dropper.
8. Measure and record the volume of fat in the cylinder.
9. Because 1 mL of fat has a mass of about 1 g and because your original sample was 100 g, the volume of fat in the graduated cylinder is the approximate percentage of fat in the ground beef.
10. Record your data and repeat the experiment with another ground beef sample.

Questions

1. What effect does heat have on fat?
2. What effect does heat have on protein?
3. What assumptions must you make if you read the volume of fat as the percentage of fat in the 100-g sample of ground beef?
4. What property of fat makes this activity possible?

Notes for the Teacher

BACKGROUND

This activity is a simple way to give students experience in making measurements and recording and interpreting data. Ground beef (hamburger), lean ground beef, and ground chuck all contain different amounts of fat. The average amount of fat is usually stamped on the package. In this activity we will simply boil the meat to separate the fat. It will float to the top of the liquid where it can be easily removed.

TEACHING TIPS

1. Samples of 100 g are used to make the percentage calculation easier. However, any mass can be used.

$$\frac{\text{grams of fat}}{\text{grams of sample}} \times 100 = \text{percent fat}$$

2. We read the percentage of fat directly from the cylinder if we assume that the mass of 1 g of fat is equal to 1 mL. This value is a good approximation.
3. Notice that some of the fat may solidify and be mixed with the cooked meat. Point this situation out to students as a possible source of error in the calculations. (Does it make the calculated percentage lower or higher than it should be?)
4. Have students record their data on the board to allow a comparison and discussion of class results.
5. Do not allow the students to eat any of the food in this activity.

ANSWERS TO THE QUESTIONS

1. It melts the fat and separates it from the meat.
2. It coagulates the protein (cooks it).
3. We assume that 1 mL of fat has a mass of l g.
4. Fat is not soluble in water; therefore, it separates and can easily be poured off.

86. Are Your French Fries Greasy?

You have probably noticed that paper in which oily or greasy foods have been wrapped has transparent grease spots. You might obtain the same effect if you drop French fries on your homework. This spot is actually used as a simple test for the presence of fat. In this experiment you will extract the fat from a food sample by dissolving the fat in a fat-soluble solvent. The technique will also be used to compare the amount of fat in several different samples of French fries.

Materials

1. Plain paper, two sheets.
2. Solvent such as trichlorotrifluoroethane (TTE).
3. Vegetable oil.
4. Several brands of French fried potatoes, cooked.
5. Small test tube.
6. Conical flask (or any glass container), 250 mL.
7. Graduated cylinder, 25 or 50 mL.
8. Balance.

Procedure

Part A: Identify the fat.

1. Select about six French fries and place them on one sheet of paper.
2. Observe the appearance of the paper as a result of the fat.
3. Obtain a 2-mL sample of the solvent you will use in a small test tube.
4. Add 2 mL of water to the test tube. Observe. Do the liquids mix?
5. Add a dropper full of vegetable oil. Does it dissolve in the water?
6. Cover and carefully shake the tube. Does the vegetable oil dissolve in the solvent?
7. Carefully record all observations from Procedures 2–6. Pour contents of tubes into the waste jar in the fume hood.

Part B: Determine the amount of fat.

8. Prepare a data table to record the masses before and after the extraction, your calculated mass difference, and your calculated percentage of fat in the sample.
9. Find the mass of the sample of French fries.
10. Place the French fries in the flask. In the fume hood, add just enough solvent to cover the French fries, about 25 mL. Stopper the flask. Avoid breathing the vapors.

11. Swirl the flask for several minutes to dissolve fat in the solvent.
12. Remove 2 drops of solvent from the flask to a piece of paper. Place 2 drops of pure solvent next to these drops.
13. Compare these spots before and after evaporation.
14. Pour off the solvent into the waste jar in the fume hood.
15. Place the French fries on a piece of paper in the fume hood so that the rest of the solvent can evaporate.
16. Determine the mass of the French fries. Why is the mass less than the original mass?
17. Determine the amount of fat that dissolved in the solvent and the percentage of fat in your sample of French fries.

Reaction

The organic solvent is only slightly polar because the electrons of each molecule are distributed almost evenly. A water molecule, however, is polar because the electrons are frequently more concentrated at one end of the molecule than at the other. Thus, one end is a little bit negative, and the other end is a little bit positive. The slightly polar and polar molecules will not mix. The fat is only slightly polar, like the organic solvent, so the fat will dissolve in the slightly polar solvent. Slightly polar solvents are also used to remove grease spots on clothing.

Questions

1. What mass of substance left the French fries in this activity?
2. What percentage of the French fries is assumed to be fat?
3. Compare the fat contents of the various brands of French fries. Compare the appearances of the various brands of French fries.
4. What variables of the various brands of French fries seem to be associated with the amount of fat (e.g., size, shape, surface area, and density)?
5. Describe two tests you did to find out whether or not fat dissolves in the solvent.

Notes for the Teacher

BACKGROUND

Fats and oils are used in the deep-fat frier for making French fried potatoes. Many people are concerned about the amount of fat in their diets and avoid foods such as French fries. However, the amount of fat retained on the food varies. In this activity students can study the variables responsible for these variations in the amount of fat. They can also learn some techniques of extraction and add to their knowledge as consumers. Extraction procedures are possible because of the relative solubility of the substance under study in this case, namely, fat. Students may notice many properties of less polar or nonpolar solvents, in general. These solvents will not mix with water, will dissolve fats and oils, and most will evaporate rapidly.

TEACHING TIPS

1. The word "solvent" is used here to refer to the organic solvent, TTE. Water is also technically a solvent for solutions in which polar substances are dissolved. Other slightly polar solvents may be used. TTE is good because it is more dense than water. Thus, in Procedure 5, three layers are observed in the tube. Upon shaking, the top oil layer will dissolve in the lower TTE layer. Then, two layers will be observed with water on top of the TTE.
2. Many organic solvents are flammable. Be sure that no flames are present. Good ventilation, as in a fume hood, is necessary. TTE is toxic to breathe.
3. Any salt on the French fries will not dissolve in the solvent. However, have students try to remove all material from the flask during the final drying period.
4. Some students may want to repeat the extraction to be sure that all of the fat is removed. Compare masses before and after.
5. This activity is ideal for allowing students some practice with the preparation and use of a data table.
6. The more brands available, the more interesting this activity is. Students should be able to bring in French fries from as many sources (including the school cafeteria) as you have student teams. The French fries can be frozen in plastic bags until ready for use.

7. You might want to prepare some French fries at school to include in the study.
8. The most significant variable will be surface area.
9. You might discuss the assumption that all mass lost by the French fries is fat.
10. Do not allow students to consume any materials in the laboratory.

ANSWERS TO THE QUESTIONS

1. Subtract final mass from initial mass. One gram of fat in the diet provides 9 kcal of energy. One kcal is equal to one nutritionist's Calorie.
2. The percentage will be small.
3. Answers will vary.
4. Answers will vary.
5. Oil in the test tube dissolved in the solvent. Fat from the French fries dissolved and remained on the paper after evaporation.

87. Bananas: The Riper They Are, the Sweeter They Are

The banana consists of approximately 75% water and 25% carbohydrate and only a trace of protein and fat. It also has quite a bit of calcium and phosphorus and is especially rich in potassium. The carbohydrate in the typical yellow banana is mostly in the form of starch. However, as the banana ages and ripens, the starch is converted to sugar, which gives the ripe fruit a very sweet taste. In this activity we will test the unripe and ripe banana for starch and sugar and compare the results.

Materials

1. Two bananas: one yellow-green (unripe) and one dark (ripe).
2. Soluble starch (cornstarch or laundry starch).
3. Clear syrup (Karo or similar brand).
4. Iodine solution.
5. Fehling's solution.
6. Dropper.
7. Test tubes.
8. Burner.

Procedure

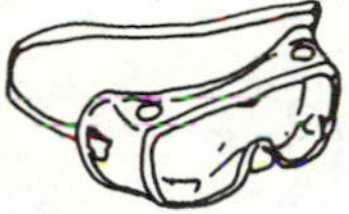

1. Test for starch.
   A. Make a paste of soluble starch in about 10 mL of warm water.
   B. Add 1 drop of iodine solution to the starch paste. Notice the dark blue formed by the starch–iodine complex. This test is the characteristic test for the presence of starch.
   C. Cut a thin slice of the yellow banana and also a slice of the dark ripe banana.
   D. Add 1 drop of the iodine solution to the surface of each slice. What do you observe? Compare the results with your test of the starch solution in step B. Which of the two banana samples contains starch?
2. Test for simpler sugars (glucose).
   A. Place about 2 mL of clear syrup in a test tube.
   B. Add about 10 mL of Fehling's solution, and shake the tube to mix the two. Place the tube in a beaker of boiling water for 10–15 min. What do you observe? The following chart will give you a general idea of the amount of sugar in the sample:
      0.5% glucose: Green produced.
      1.0% glucose: Yellow produced.
      2.0% or more glucose: Orange-red produced.
      This test is the characteristic test for a simple sugar.
   C. Place a slice of the yellow banana in a test tube. Place a similar sample of the dark ripe banana in another test tube.
   D. Add 10 mL of Fehling's solution to each tube, mash the banana with a stirring rod, and place the tubes in a beaker of boiling water. What do you observe? Which sample contains the most simple sugar?

Reaction

Starch is the primary carbohydrate in the unripe banana. Starch has a very complex structure, but when it reacts with water, a process called *hydrolysis*, it breaks up into smaller molecules called *dextrins*. Upon further treatment, these dextrins break up into molecules of glucose, which is a simple sugar:

$$\text{starch} \xrightarrow{H_2O} \text{dextrins} \xrightarrow{H_2O} \text{glucose}$$

Starch in the unripe banana gives a positive test (blue) with iodine. About 20% of starch is amylose, which is a chain of glucose molecules bonded together. The chain is twisted into a "helix", like a spring. The inside of the helix is just the right size to accept iodine. The iodine is also complexed, I_5^-. The iodine lodged inside the helix forms the deep blue starch–iodine complex. Dextrins will produce brown with iodine solution. Glucose does not produce a color with iodine solution.

Glucose is one of the sugars that gives a positive test with Fehling's solution. Only the ripe banana gives a positive test. This result indicates that the starch has been converted to glucose.

$$\text{glucose} \xrightarrow{\text{Fehling's solution}} \text{red precipitate}$$

Questions

1. Describe the positive test for starch.
2. Which banana sample gave a positive starch test?
3. Describe the positive test for simple sugar.
4. Which banana sample gave a positive test for a simple sugar (glucose)?
5. Test some other foods to see if they consist of starch or sugars.

Notes for the Teacher

BACKGROUND

This activity gives students experience with qualitative analysis. They will first establish a positive test for starch and glucose. Then, they will test an unripe banana and a ripe (dark) banana for starch and glucose. The starch in the yellow unripe banana is slowly converted to glucose as the banana darkens. They can also test other samples of foods for these two carbohydrates.

SOLUTIONS

1. Clear syrup solution: Dilute the syrup 50:50 with water.
2. Fehling's solution. Prepare this solution as follows:
   A. Make solution A by dissolving 87 g of sodium citrate and 50 g of crystalline sodium carbonate in 425 mL of warm water.

B. Make solution B by dissolving 8.7 g of copper sulfate pentahydrate in 50 mL of water.

C. While stirring constantly, pour solution B into solution A. Dilute to 500 mL.

3. Potassium iodide–iodine solution: Either dilute tincture of iodine from the drugstore with water or make the solution as follows:

A. Dissolve 2.5 g of potassium iodide in 500 mL of water.

B. Add 1.0 g of iodine crystals and stir until they dissolve. Place the solution in small dropper bottles.

TEACHING TIPS

1. Starch has no sweet taste. Glucose is sweet, but not as sweet as table sugar.
2. Sucrose (table sugar) is not produced from starch. It also does not give a positive test with Fehling's solution.
3. A water bath is used to provide even heating of the solutions.
4. Instead of Fehling's solution, you can also use diabetic test sticks or tablets from the drugstore to test directly for glucose, or you may use Benedict's solution.
5. A *tincture* is an alcohol solution. Tincture of iodine or potassium iodide–iodine solution is also called *Lugol's solution*.
6. Be sure that students heat all solutions long enough to produce the end color.
7. Test various foods (e.g., potatoes, crackers, and fruit) for sugar and starch content.
8. Fehling's reagent will give a positive test for any sugar with a free aldehyde group. Such a sugar is called a *reducing sugar*. Glucose, galactose, fructose, and maltose are all reducing sugars. Sucrose is NOT a reducing sugar. Reducing sugars reduce the cupric ion in Fehling's solution to the red cuprous oxide, which precipitates. The reaction is essentially the following:

$$\underset{\text{blue}}{2Cu^{2+}(\text{citrate})} + R\overset{\overset{O}{\|}}{C}H + 5OH^- \longrightarrow R\overset{\overset{O}{/\!/}}{C}\!-\!O^- + \underset{\text{red-orange}}{Cu_2O(s)} + 2\text{citrate}^{2-} + 3H_2O$$

9. The blue color is produced when the I_5^- or I_3^- complex fits inside the coiled molecule of amylose, one of the two components of starch.

ANSWERS TO THE QUESTIONS

1. Starch produces dark blue–black with iodine.
2. The unripe banana gave a positive starch test; the ripe banana did not give a positive starch test.
3. Glucose (and some other simple sugars) gives a red precipitate when heated with Fehling's solution.
4. The dark ripe banana gave a positive test for glucose.
5.

Starch Positive Test	*Simple Sugar Positive Test*
crackers	apple juice
potatoes	fruit
bread	toasted bread

Chemical Detectives: Tools and Techniques of the Chemist

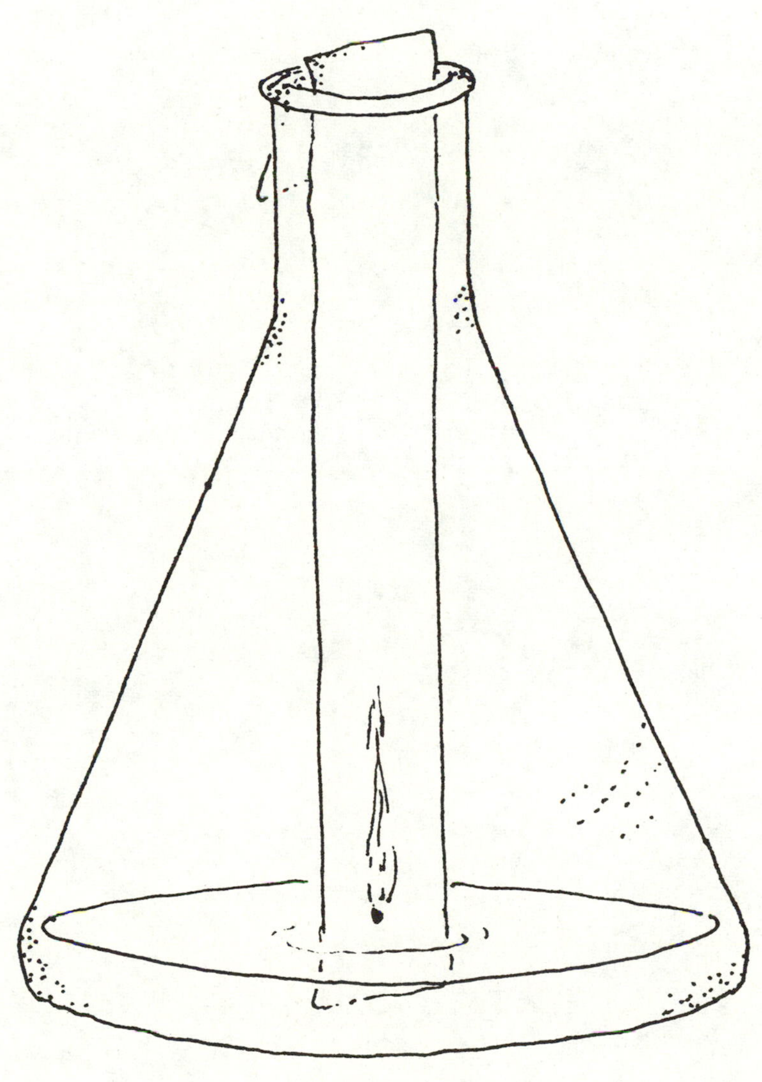

88. Paper Chromatography

In this activity we will use a simple technique to separate the colors from black ink. Because the ink is soluble in water, we will allow the solution to be absorbed on filter paper. As the solvent rises on the paper and evaporates, it leaves the colors in its path.

Materials

1. Ordinary filter paper or chromatography paper.
2. Black felt-tip pens (Flair pens and others).
3. Small jar, bottle, or flask.

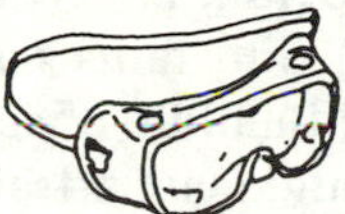

Procedure

1. Cut a strip of filter paper long enough to fit easily almost to the bottom of the container and to fold about 1 cm over the lip. The strip should be about 1 in. (2.5 cm) wide.
2. Using a black felt-tip pen, place a small dot about 0.5 in. (1.3 cm) from one end of the filter paper.
3. Add enough water to the container to cover the bottom of the filter paper but not enough to reach the ink dot.

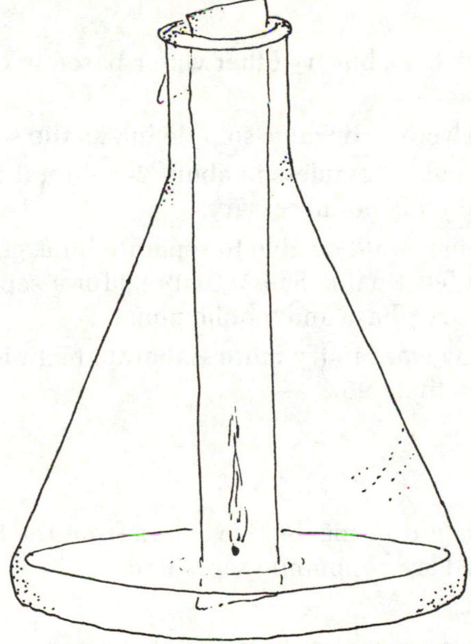

4. Place the filter paper in the container with the dot end facing down.
5. Observe.

Reaction

Because the ink used in this activity is water-soluble, it dissolves and is adsorbed on the filter paper. As the ink rises on the paper, the different colors are deposited according to their degree of solubility in water and their attraction to the paper. The more soluble the color, the farther these pigment molecules tend to move up the paper.

Questions

1. What colors separated from the black ink?
2. Would other colored inks separate as well? (Try some!)
3. What practical application does chromatography have?

Notes for the Teacher

BACKGROUND

Chromatography is defined as the distribution of a solute (ink) between a stationary phase (filter paper) and a mobile phase (water). Substances are separated according to their relative attraction for the stationary and mobile phases. The solute (color) with the greatest attraction for the solvent and least attraction for the paper will be the last to deposit on the paper.

This experiment is a simple activity to introduce students to the technique of separation by chromatography. If you have chromatography paper it will work well, but ordinary filter paper will work just as well. You probably will not be able to use paper napkins or towels because they absorb so strongly that color separations are not always seen (this is what they are made to do). Coffee filters work well.

TEACHING TIPS

1. Flair felt-tip pens seem to work best, but try other water-based ink pens. Permanent ink pens seem not to work.
2. This system uses water as a solvent and water-soluble ink as the sample.
3. The adsorption is quite rapid and is complete in about 2–3 min; thus, covering the beaker to prevent evaporation is not necessary.
4. Colors are nicely separated. You should be able to separate blue, green, yellow, and maybe red from the black felt-tip ink. See Activity 89 for a separation technique using a different stationary phase and mobile phase.
5. After this activity, students may want to try more elaborate techniques to separate and identify metal ions (Activity 90).

ANSWERS TO THE QUESTIONS

1. You should be able to separate four or maybe five colors from the black felt-tip ink. Blue, green, yellow, and red are commonly separated.
2. Various colors with other inks.
3. Send students to the library to research applications of chromatography. There are many types of chromatography.

89. Chalk Chromatography

Chromatography is a technique that scientists use to separate substances from a mixture. In this activity we will separate the colors in black ink. We will dissolve the ink in a suitable solvent and then allow this solution to be adsorbed on a piece of chalk. As the solution travels up the chalk, various colors will appear because each color in the ink is adsorbed on the chalk with a different strength.

Materials

1. Small beaker (100 mL).
2. Large beaker (250 or 500 mL).
3. New piece of porous white chalk.
4. Black felt-tip pen.
5. Methanol (about 25 mL).

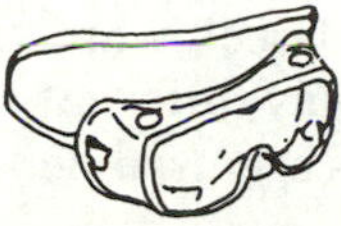

Procedure

1. Check to be sure that the piece of chalk will stand on end in the small beaker. If not, scrape the surface of the chalk until it does.
2. Using the black felt-tip pen, make a dark dot on the chalk about 1.25 cm from the flat end. Place several other dots around the chalk the same distance from the end.
3. Stand the chalk on end in the small beaker.
4. Wear gloves. In the fume hood, carefully add methanol to the small beaker until the level is just below the ink dot. DO NOT INHALE THE METHANOL VAPORS.

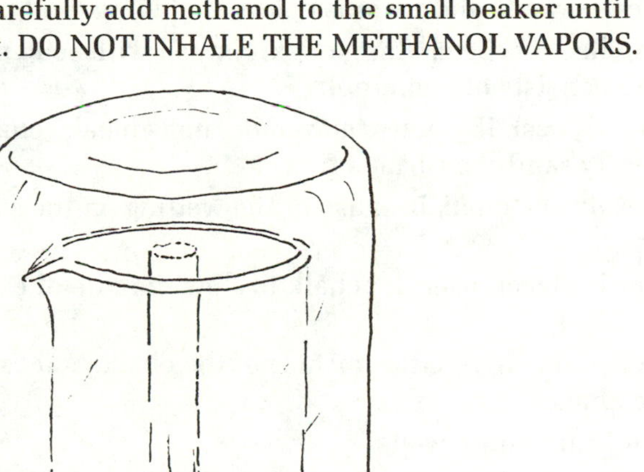

5. Invert the large beaker and use it to cover the small beaker.
6. Observe.
7. Return the methanol to the special storage container in the fume hood.

Reaction

The ink dissolves in the methanol, which is an organic solvent. The various components (colors) of the ink are adsorbed on the chalk with a different strength; thus, they separate as the solvent moves up the chalk column. The colors that travel the farthest are the most soluble in the methanol and adhere most weakly to the chalk.

Questions

1. In what order did the colors appear on the chalk?
2. What purpose did the chalk serve? What else could have been used for the same purpose?
3. What colors would separate from green ink? (Try it!).
4. What practical purpose does this procedure have?

Notes for the Teacher

BACKGROUND

The process of chromatography was first developed by a Russian botanist, Michael Tswett, in 1906. Because the process was first used only to separate colored compounds, the name of the process was derived from the Greek word "chroma", which means color. Now the process is used to separate many different components from a variety of mixtures. Also, different types of columns are used depending on the nature of the solvent and the materials to be separated.

TEACHING TIPS

SAFETY NOTE: Dispense methanol in the fume hood. It is flammable and toxic.

1. Our experience shows that the Flair felt-tip pen works best. You should experiment with various brands and various colors, including food coloring.
2. If methanol is spilled on the clothing or skin, wash with soap and water. Some duplicating fluid consists of methanol.
3. Porous chalk works best. If you use glazed or "nonsqueak" chalk, it may be necessary to lightly sand the surface first.
4. The solvent rises up the chalk because of the wetting on the surface, similar to capillary action.
5. The large beaker is placed over the chalk to slow the rate of evaporation of the solvent.
6. The adsorption occurs only on the surface of the chalk. You can easily show this by breaking the chalk.
7. Try other solvents, including water.
8. You could stand the chalk in a beaker and cover the top with a plastic wrap.
9. Keep the methanol covered. Store used methanol until it is needed again for chromatography.

ANSWERS TO THE QUESTIONS

1. You should be able to separate at least three colors from black: red, orange-yellow, and blue. You may also get green.
2. The chalk serves as a column onto which the colors can be deposited. Capillary action causes the solvent to rise in the column. You could also use paper or silica gel.
3. Green ink should produce yellow and blue.
4. Send students to the library to look for practical applications (e.g., separation of amino acids to determine protein structure and analysis of compounds).

90. Separating Metal Salts in Water by Paper Chromatography

Separating dissolved metals from each other is difficult because many have similar chemical reactions. However, a simple way is to spot the solution on a special paper and allow a solvent to pull the metals along as it soaks up the paper. This technique is *paper chromatography* and is used for separation and identification of many substances in the laboratory.

Materials

1. Any combination of the following salts in solution: iron(III) chloride ($FeCl_3$), cobalt(II) chloride ($CoCl_2$), manganese(II) chloride ($MnCl_2$), and copper(II) chloride ($CuCl_2$).
2. Separate solutions of each of the salts chosen.
3. Filter paper or chromatography paper.
4. Acetone–hydrochloric acid developing solvent (350 mL of acetone + 100 mL of 6 M HCl).
5. Capillary tube for each solution.
6. Beaker, 600 mL, or peanut butter or mayonnaise jar.
7. Plastic wrap.
8. Ammonia (NH_3) solution, 6 M.
9. Sodium sulfide (Na_2S) solution, dilute.

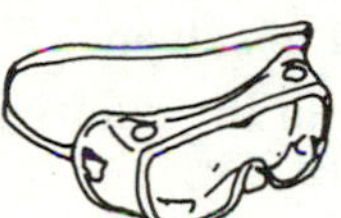

Procedure

1. Cut a rectangle of filter or chromatography paper to be at least 9 cm from top to bottom and 12 cm long.
2. With a pencil, draw a line about 1 cm from the bottom (long edge) and parallel to the bottom of the paper.
3. Place an "X" on the line every 2.5 cm.
4. Label each "X" for one of the solutions you will be using.
5. Using the capillary tube, spot a small dot of a solution on the corresponding "X".
6. Dry the spots.
7. Repeat Procedure 5, adding a second drop of the same solution on top of the first.
8. Form the paper into a cylinder and secure the ends together with two staples. Do not overlap the paper.
9. Wearing gloves, goggles, and a face shield, place about 40 mL of the developing solvent in the beaker. Be sure the level of the solvent remains below the spots.
10. Place the paper cylinder carefully into the solvent with the spots at the bottom. Be sure that the solvent does not touch the spots.
11. Cover with plastic wrap and observe.
12. When the solvent has almost reached the top of the paper, remove the paper and place a pencil mark where the solvent stopped.
13. Let the paper dry and observe any spots.
14. Wear gloves, goggles, and a face shield. In the fume hood, spray the paper with 6 M ammonia and observe any spots.
15. Using a dropper full of sodium sulfide solution, streak the dropper across the paper up from each "X" to the top of the paper and observe the spots.
16. Compare the known dissolved metal with any unknowns to identify the metals.

Reactions

Four types of reactions are occurring: (1) formation of chloride complexes when the spot moves up the paper, (2) formation of ammonia complex, (3) formation of sulfide, and (4) relative solubility of the metal salt in the solvent. Iron(III) ions form a yellow chloride complex. Cobalt forms a blue chloride complex. Copper forms a yellow complex. The manganese complex is pale pink. Copper ions form a blue ammonia complex. Cobalt forms a pinkish ammonia complex. With ammonia, iron is visible as brown iron(III) hydroxide, $Fe(OH)_3$. The sulfides of each of the metals show up well as a spot of the black, brown, or peach (manganese) sulfide. For example:

$$Mn^{2+}\,(aq) + S^{2-}(aq) \longrightarrow \underset{\text{peach}}{MnS\,(s)}$$

The most soluble dissolved metal salt moves the farthest with the solvent. The solubility is greatest for iron and then for copper, cobalt, and manganese.

Questions

1. Which dissolved metals had the greatest attraction for the paper?
2. Which dissolved metal had the greatest solubility?
3. What is the ratio of the distance traveled in millimeters of the dissolved copper compared with the distance traveled by the solvent?
4. Why do you think the colored chloride and the colored ammonia complexes faded?
5. How can you identify an unknown dissolved metal salt with this technique?

Notes for the Teacher

BACKGROUND

Paper is made of cellulose, which is a complex polysaccharide molecule having many places that attract charged metal ions such as iron, copper, cobalt, and manganese. The solvent, acetone, also attracts the metal ions. As the solvent moves up the paper, carrying the ions with it, a "tug of war" occurs between the paper and the solvent. Once the solvent evaporates, the metal ions stay on the paper, but each of the ions stays at a different point. The result is that the metal ions are separated from each other. Paper chromatography is a powerful tool used in analytical laboratories for the separation of many kinds of substances including biochemicals such as amino acids.

SOLUTIONS

1. Prepare the metal salt solutions by placing about 1 tsp of the salt in 100 mL of 1 M HCl.
2. The solvent is made by adding 100 mL of 6 M HCl to 350 mL of acetone. The acid is corrosive. Wear gloves, goggles, and a face shield.
3. Sodium sulfide (Na_2S) solution is prepared by adding about 1 tsp of Na_2S to 100 mL of water.

TEACHING TIPS

1. Nitrate salts may be substituted for the chlorides.
2. Varying the amount of HCl may be a good investigation for an interested student.
3. Whatman No. 1 chromatography or filter paper works well. You may try others.
4. If you do not have capillary tubes, try a piece of straw. The spot of solution should be small.
5. The paper should be dried between each application of solution. A hair dryer is a useful tool.
6. For Procedure 14, if you have mayonnaise jars and a hood, try pouring a few milliliters of concentrated ammonia on several cotton balls in each jar. Students may add their papers and screw on the lids. Then they can take their jars to their laboratory benches and watch the spots develop as the ammonia complexes form. Be sure that the lids are not removed outside of the hood. Household ammonia can also be used. It will take longer.
7. If you place sodium sulfide solution in dropper bottles, the students can draw the tip of the dropper along the path of the ions from each "X" to discover the position of the spot as the sulfide of the metal forms.
8. This activity is especially meaningful if you prepare unknowns for the students of single or mixed metallic ions.
9. You may also want to try zinc ions to show that not all transition metals form colored salts or colored complexes.
10. You may want to teach students to find the retention factor (Rf), which is the ratio of ion distance traveled to solvent distance traveled. Measure, in millimeters, to the most dense part of the final spots from the "X".

ANSWERS TO THE QUESTIONS

1. Cobalt and manganese.
2. Iron.
3. About 0.75.
4. The HCl and the NH_3 vaporized.
5. Identify its reactions and the distance it traveled.

91. Colored Flames

When atoms of elements are heated or electrified, some give off visible colored light. For example, the familiar red glow of a neon sign is caused by neon atoms excited by electricity. Fireworks displays of bright lavender, red, green, and yellow are the result of excited metal atoms. In fact, the kind of color given off by each atom is so exact that it can be used to identify that atom. You will test various metal salt solutions in a hot flame to learn the expected color given off by each excited atom.

Materials per Team

1. Bunsen burner or propane torch (in cradle).
2. Wooden splints (soaked in salt solutions) or nichrome wire in wooden handles.
3. Tongs.
4. Solutions: lithium chloride (LiCl), barium chloride ($BaCl_2$), sodium chloride (NaCl), strontium chloride ($SrCl_2$), potassium chloride (KCl), and calcium chloride ($CaCl_2$).

Procedure

1. Remove a piece of soaked wooden splint from the solution with forceps or tongs.
2. Hold the soaked splint in the flame of a burner. Record the color.
3. Repeat with each salt solution.
4. Obtain a soaked splint with one or two salts on it that are unknown to you. Test them in the flame and report to your teacher what metal atoms you believe were on your splint.

Reaction

The vaporization and glowing of these metal atoms are not chemical reactions. The electrons in each atom are promoted to a higher energy by the heat. When these electrons fall back to their original level, they emit the same energy in the form of light instead of heat. You give the electrons heat so that they can jump up, and they give you back light when they fall back. These atoms give you beautiful colored light.

Questions

1. If you have some road salt, how can you tell if it is calcium chloride or sodium chloride?
2. If you could design a fireworks display, what salts would you use and why?

Notes for the Teacher

BACKGROUND

The identification of elements was made much easier after scientists found that each element showed a characteristic spectrum when it glowed. The secret was to get the atoms hot enough. This need was satisfied by Robert Bunsen when he developed his famous burner in 1855. He is even more renowned, however, for his development of the *spectroscope*, which separates wavelengths of light from each other so that a particular pattern of bright lines can be seen when the atoms glow. Bunsen discovered the elements cesium and rubidium in 1860–61. Notice sometime that when table salt spills near a hot stove, it glows yellow. Try salt substitute. The potassium glows lavender.

SOLUTIONS

Prepare saturated solutions of each salt. A solution is saturated if the maximum amount of salt is dissolved. If you do not have the chlorides of barium, strontium, or calcium, use any compound that you have. However, the best results are obtained if the chlorides can be used. If your salt is not a chloride, add 6 M HCl to the salt to make a concentrated metal chloride solution to use with the splints. Caution students to be careful with the acid or prepare the chlorides for them.

Place enough wooden splints in each container of solution to provide a splint for each student. Allow the splints to soak overnight.

TEACHING TIPS

1. The flame colors are short-lived and of varying intensity. Students will sharpen their powers of observation. The potassium flame is difficult to detect because it is obscured by the yellow flame. Look through two thicknesses of cobalt blue glass to mask the yellow. The potassium flame can still be seen. Didymium glasses, used by glass blowers and welders, make an excellent alternative to the cobalt. They are fun to use and can be obtained from a welders' supply house for about $20.00.
2. Flame colors: lithium, red; sodium, yellow; potassium, lavender; calcium, red-orange; barium, green; and strontium, scarlet-red.
3. Mercury vapor lamps, halide vapor headlights, and sodium vapor lamps all depend on glowing atoms in the same way as the neon sign does.
4. This activity requires a hot flame. The flame of a Bunsen burner is 800–900 °C.
5. Instead of soaked wood splints, you may use wire loops. Straighten a paper clip and make a tiny loop at one end. Soaked filter paper held with tongs also works well.

ANSWERS TO THE QUESTIONS

1. Use flame tests.
2. Have students do a little library research on fireworks.

92. Distillation of Vinegar

Water is a pure liquid. When a liquid is not a pure liquid, we can sometimes separate each part by a process called *distillation.* Chemists use this process to purify solvents and samples of substances that have been prepared in their laboratories. Crude oil is distilled to separate each part from butane to gasoline to paraffin wax. Distillation is also the process used in the beverage industry to separate ethyl alcohol from the mixture in which it was produced by yeast. Near the ocean, distillation can be used to separate pure water from the mixture of salts that make up sea water.

Materials

1. Hard glass test tube or small flask (e.g., Pyrex).
2. Stopper with two holes to fit boiling tube or flask.
3. Thermometer for stopper.
4. Glass bend to fit hole in stopper.
5. Rubber tubing (about 6 in.) to fit glass bend.
6. Burner.
7. Collecting test tube.
8. Beaker for cooling water bath.
9. Boiling chips or piece of porcelain.
10. Vinegar.

Procedure

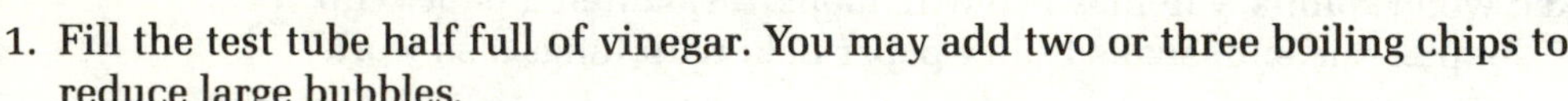

1. Fill the test tube half full of vinegar. You may add two or three boiling chips to reduce large bubbles.
2. Insert the stopper with thermometer and glass tube. Be sure that the thermometer is above the liquid. (Why?)
3. Place the rubber tubing in the collecting test tube in a beaker of cold water.

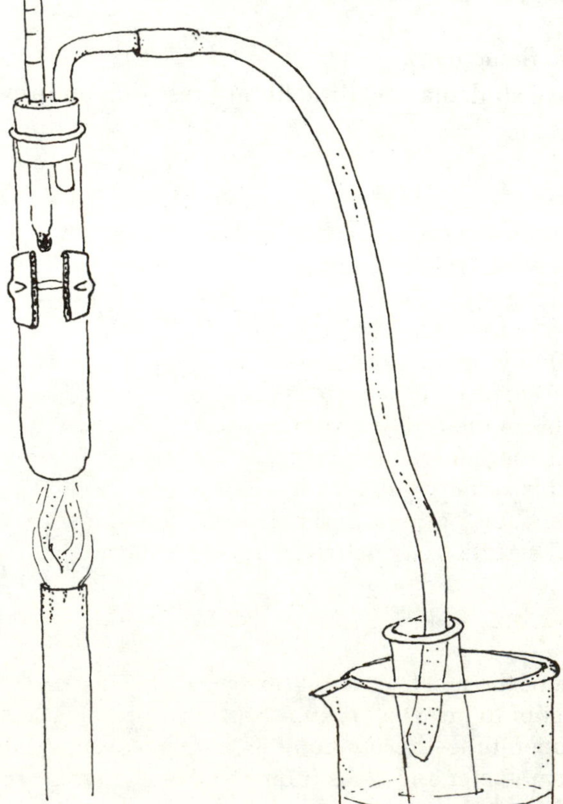

4. Gently heat the vinegar.
5. Record the temperature every 30 s. Keep the heat even.
6. Make a graph of time (*x* axis) versus temperature (*y* axis) as the liquid heats and boils.

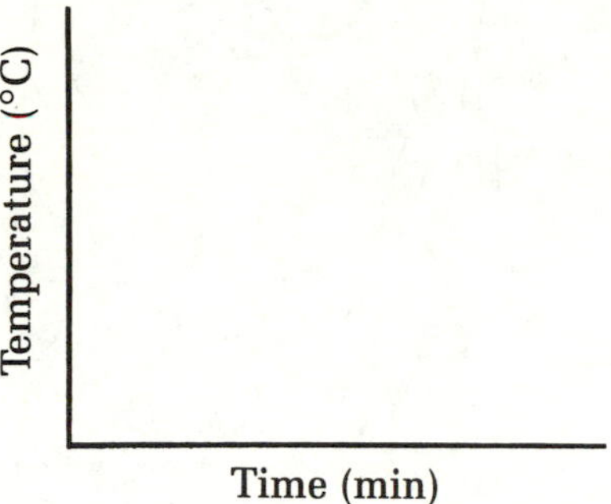

Reaction

The boiling point of a pure substance is constant over time as the substance boils and becomes a gas. However, vinegar is a mixture of acetic acid, CH_3COOH, and water. Your solution will be boiling and becoming more concentrated as the water becomes a gas. You may constantly see a temperature change. The boiling point of pure water is 100 °C. The boiling point of pure acetic acid is 118 °C.

Questions

1. Describe what you see during the boiling process.
2. What do you think is inside the bubbles that rise to the surface?
3. Describe the shape of the curve you drew on your graph.
4. Explain why the curve changed shape as it did. What was happening to the molecules in the liquid?

Notes for the Teacher

BACKGROUND

Distillation is an important separation technique used in the analytical laboratory. The petroleum products we use are separated from crude oil by fractional distillation. Each fraction in a mixture of liquids has a different boiling range. When the boiling range of a fraction occurs, the molecules are absorbing the energy being supplied as heat. The molecules use the energy to move more rapidly. As they move more rapidly, some molecules overcome the attractions to each other and escape from the surface of the liquid. These molecules, then, are gas molecules. When the molecules in the first fraction have left the liquid, the liquid increases in temperature and another fraction begins to boil, that is, it becomes gaseous.

In this activity you may have the students distill a variety of mixtures, for example, sea water or any salt solution, the ingredients in mouthwash, or the solution that results from Activity 73, "Fermentation and Apple Cider".

TEACHING TIPS

1. One of the most common laboratory accidents involves cuts when students attempt to insert glass tubing into rubber stoppers. Prepare the rubber stopper–tube–thermometer assembly for the students. Place glycerin on the thermometer and glass tubing and gently insert them into the stopper with a slight twisting motion. Be sure that the end of the glass tubing is fire-polished. Hold the stopper with a cloth towel.

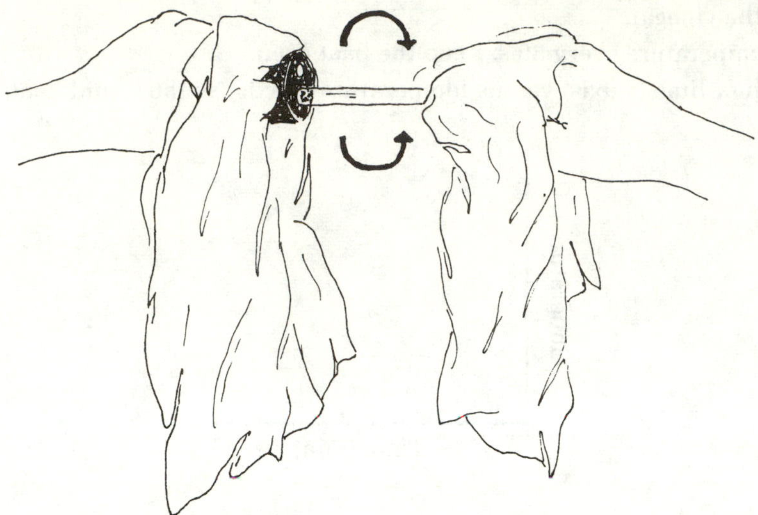

2. A dramatic first distillation is the separation of water from a dilute solution of potassium permanganate (purple).
3. A worthwhile preliminary activity to this one is to heat and boil pure water while taking time and temperature data. These data can be used to practice graphing.
4. If you distill the fermented cider, you will need to watch closely for the temperature to level off somewhat at 78 °C (the boiling point of ethyl alcohol). The percentage of ethyl alcohol in the cider will be low. Alcohol and water form an *azeotrope*, which means that some molecules of one remain associated with the molecules of the other, even when changing from gas to liquid. Azeotropes do not completely separate with this procedure. Their boiling points are slightly different from either of the two liquids.
5. The substances that are separated by distillation from petroleum include the following (boiling points are given in parentheses): methane (−161 °C), propane (−43 °C), octane (125 °C), kerosene (200 °C), petroleum jelly (475 °C), and solid residue (asphalt for roofing and roads).

ANSWERS TO THE QUESTIONS

1. Answers will vary but should include the formation of small and then larger bubbles that rise to the top of the liquid.
2. Vapor of the liquid.
3. The temperature increase resulted in a slanted rising line that formed a flat plateau while the liquid boiled.
4. As the molecules gained energy to escape the liquid (boiled), the temperature of the liquid stayed the same.

93. Distillation of Wood

It might surprise you to find that heating wood in an enclosed test tube will not cause it to burn; however, it will cause the wood to decompose to form gases and liquids and leave a black solid residue.

Materials

1. Wood splints (Popsicle sticks work well).
2. Two hard glass test tubes, 18 × 150 mm.
3. Rubber stopper with glass bend to fit.
4. Burner.
5. Beaker of cold water.

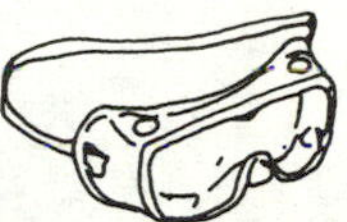

Procedure

1. Loosely pack test tube 1 with wood splints.
2. Attach the rubber stopper with the glass bend to the test tube.
3. Place the end of the glass bend into test tube 2, and place test tube 2 in the beaker of cold water.

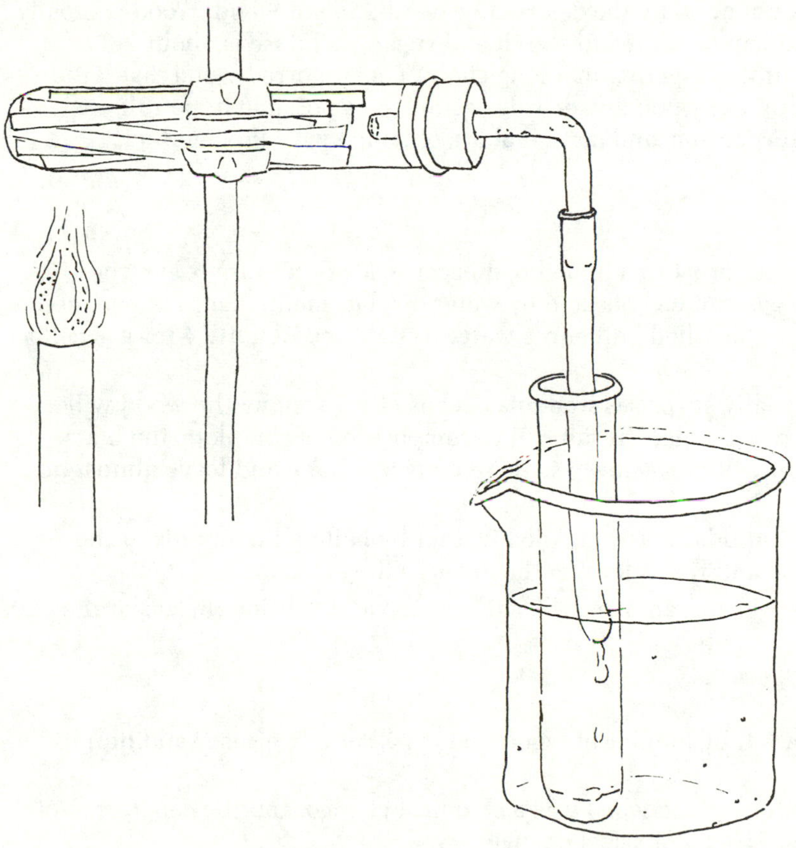

4. Heat tube 1, which holds the wood splints. (You may hold the tube with a clamp, or use a heat-proof clamp on the ring stand.)
5. Observe the liquids condensing in cooled test tube 2.
6. Observe the gases that are evolved. How could you collect the gases?

Reaction

Wood is mostly cellulose, which decomposes when heated in the absence of air. A mixture of liquids and gases are formed including turpentine, methanol (wood alcohol), and methane gas.

Questions

1. Could you tell by looking at wood that liquids and gases would be obtained by heating?
2. Was this a chemical or physical change?
3. What is the difference between heating wood in air and heating wood without air?
4. Describe the appearances and odors of the products.

Notes for the Teacher

BACKGROUND

Strong heat can decompose wood and then separate the products by vaporizing them, after which several products are condensed in a cooled tube. Commercially, many products are obtained by the destructive distillation of wood. Wood is mostly *cellulose*, a large molecule related to starch and sugar. Cellulose is about 2800 glucose molecules linked together in a long chain. The preferred wood distillation process is to grind up the wood and dissolve as much as possible in petroleum solvents. Pine oil, turpentine, and methanol are obtained with this method.

TEACHING TIPS

1. You may do Procedure 4 in a fume hood because a gas mixture is evolved. Otherwise, the gas can be collected by water displacement. Place rubber tubing in an inverted water-filled bottle in a water trough. See Activity 4 for gas collection procedure.
2. This activity usually surprises students because they assume the wood will burn. Try burning the black residue. If decomposition is complete, the black residue (carbon) will burn slowly with an invisible flame and leave almost no ash.
3. The condensed liquids can be distilled further by boiling the liquids in the condensing tube and recondensing the lower boiling fraction.
4. Try burning the gas by removing the rubber tube and lighting the gas at the end.

ANSWERS TO THE QUESTIONS

1. No, you cannot tell by looking at wood that it will produce gases and liquids when heated.
2. The decomposition of the wood was a chemical change, and the separation of liquids and gases was a physical change.
3. Heating wood in air causes it to burn to form the products carbon dioxide and water.
4. Three types of products can be seen. The wood becomes a black solid residue (charcoal). Liquids condensed in the cooled test tube are dark yellow-brown. Gases and some smoke are also produced. The gas has a strong odor.

94. Charcoal: A Chemical That Eats Odors and Colors

In this activity we will remove color from a solution and odor from another solution by allowing the solutions to react with carbon. We will use carbon that has undergone a special treatment to make it into *decolorizing* carbon. Many substances can be purified by allowing them to react with this carbon. Objectionable taste and odor can be removed from our drinking water, and colors can be removed from many types of solutions.

Materials

1. Funnel.
2. Filter paper.
3. Decolorizing or activated carbon.
4. Black ink or food colors.
5. Potassium permanganate ($KMnO_4$) solution.
6. Sauerkraut or pickle juice.

Procedure

1. Fold a piece of filter paper, place it in a funnel, and put the stem of the funnel into a large test tube in a test tube rack or into another beaker.
2. Add about 1 tsp of decolorizing carbon to the funnel.
3. Add 1 drop of ink or food color to 100 mL of water in a beaker.
4. Carefully pour some of the colored water onto the charcoal in the filter paper. Describe the liquid filtered into the test tube (the *filtrate*).
5. Discard the filter paper and charcoal in the solid waste receptacle.
6. Prepare another filter paper with the same amount of carbon. This time filter a solution made by adding two or three small crystals of potassium permanganate to 100 mL of water in a beaker.
7. Describe the liquid filtered into the test tube.
8. Repeat this activity using sauerkraut juice or dill pickle juice as the solution that is filtered. Is any odor present in the filtrate (or less odor than in the original juice)?

Reaction

Charcoal acts as a decolorizing agent and as a deodorizing agent because it holds onto molecules of chemicals that are responsible for colors and odors. This process, called *adsorption*, requires a large surface area. You can think of each individual piece of carbon as being like a sponge with many tiny holes. This porous structure gives carbon a very large surface that can attract and hold molecules of the substance being adsorbed. Gases as well as liquids can be adsorbed on the surface of charcoal. Charcoal probably reacts chemically with potassium permanganate, reducing the purple Mn^{7+} ion to the colorless Mn^{2+} ion.

Questions

1. Describe the material before and after filtration in each of the three activities.
2. How does carbon remove color from solution?
3. How could this process be used to provide pure water for drinking?

Notes for the Teacher

BACKGROUND

Charcoal is made by heating wood to a very high temperature in the absence of air. When it is heated to an even higher temperature, to about 930 °C, impurities are driven from its surface and it becomes *activated* charcoal, sometimes called *decolorizing* coal. This activated charcoal can remove impurities in either the gaseous or liquid state from many solutions. It does so by the process of *adsorption*, or by attracting these molecules to its surface. In this activity students can remove the color of a solution (ink or food coloring), the purple of potassium permanganate, and the odor from sauerkraut or dill pickle juice.

SOLUTIONS AND MATERIALS

1. Activated carbon: This carbon is a very fine powder. Students must be careful not to spill it.
2. Potassium permanganate ($KMnO_4$): Only a few crystals will be needed to prepare a deeply colored solution of this compound.
3. Sauerkraut juice or dill pickle juice: Obtain these from a grocery store. Students can bring samples from home.

TEACHING TIPS

1. Adsorption is often used to purify drinking water.
2. Adsorption on carbon is prevented if the pH of the solution is greater than 9 (quite basic).
3. Adsorption by charcoal is also used to remove unburned hydrocarbons from automobile exhaust, harmful gases from the air, and unwanted colors from certain products.
4. Students may find the difference between adsorption and absorption confusing.
   ADsorption: A gas, liquid, or dissolved substance is gathered on the surface of another substance (e.g., charcoal).
   ABsorption: A liquid is soaked up, as with a blotter. It is taken in completely and mixes with the absorbing material (e.g., absorbent cotton).
5. This activity gives students experience with filtration. They could also add the charcoal to the liquid in a flask, shake the flask, and let the charcoal settle. This process would also produce a colorless solution.
6. Try removing color, odor, or taste from other solutions.
7. You might also try adding a small amount of charcoal to a colored solution, heating it to a boil, and then filtering.
8. Carbon reduces the purple Mn^{7+} ion in permanganate to the colorless Mn^{2+} ion.

ANSWERS TO THE QUESTIONS

1. In each case, the filtrate is clear and does not have the impurity (i.e., color, taste, or odor) of the original solution.
2. Charcoal has many small holes. This feature gives it a large surface area. This large surface area allows it to attract a large number of molecules of the impure substances.
3. Water is filtered through charcoal to remove impurities that would otherwise discolor and give a bad taste to drinking water.

95. Chemistry of Invisible Inks

Mystery novels and old spy movies often center around secret messages written in invisible inks that become visible only when treated in the proper manner. Although invisible inks are no longer a fad, they are fun to make and involve some interesting chemistry. In this activity you will learn several ways to write secret messages with invisible inks and the chemistry that allows you to make them visible.

Materials

1. Notebook paper.
2. Lemon juice.
3. Cobalt chloride ($CoCl_2$) solution.
4. Potassium thiocyanate (KSCN) solution.
5. Ferric chloride ($FeCl_3$) solution.
6. Cotton swabs or small art brushes.
7. Burner.
8. Several small beakers.
9. Spray bottles.

Procedure

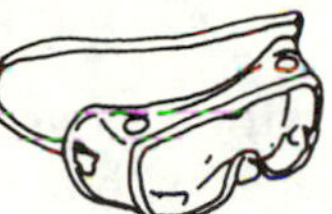

1. Invisible inks that respond to heat.
   A. Using a clean brush or a cotton swab, write a message on notebook paper with lemon juice. Write another message on a separate sheet of paper with cobalt chloride solution. Wear gloves. Cobalt chloride is toxic.
   B. Allow the papers to dry. Are the written messages invisible?
   C. Holding one end of a paper in each hand, slowly pass the paper above a hot plate or light bulb. What do you observe?
2. Invisible ink that requires chemical treatment.
   A. Write a message with potassium thiocyanate solution.
   B. Allow the paper to dry.
   C. Tape the paper to a large piece of cardboard and spray it with ferric chloride solution. What do you observe?

Reactions

1. When hot citric acid from lemon is concentrated, it chars paper by reacting with the cellulose to produce black carbon. Heating concentrates the acid and produces this reaction:

$$\text{cellulose} \xrightarrow[\text{heat}]{\text{citric acid}} C(s) + H_2O(g)$$

Cobalt chloride is light pink, almost colorless, when in solution. However, when water molecules are removed by heat, the dry, or anhydrous, blue form of cobalt chloride is formed. This form makes the written message appear blue:

$$\underset{\text{pink}}{2[Co(H_2O)_6]Cl_2} \xrightarrow{\text{heat}} \underset{\text{blue}}{Co[CoCl_4]} + 12H_2O$$

Notice the removal of water molecules from the pink cobalt chloride structure to produce the blue structure.

2. Potassium thiocyanate and ferric chloride.

   The ferric ion (Fe^{3+}) reacts with the thiocyanate ion (SCN^-) to produce the dark-red complex, $Fe(SCN)^{2+}$:

$$\underset{\text{pale yellow}}{Fe^{3+}(aq)} + \underset{\text{colorless}}{SCN^-(aq)} \longrightarrow \underset{\text{red}}{Fe(SCN)^{2+}}$$

Questions

1. Describe the appearance of the inks when they are made visible.
2. How does heat make certain invisible inks visible?
3. Describe the chemical reactions that produce visible writing from some invisible inks.

Notes for the Teacher

BACKGROUND

Invisible inks are chemicals that undergo a color change when treated with heat or with chemical solutions. In this activity students can write secret messages with a variety of substances that can be considered invisible inks. Some reveal their message when slightly heated; others require chemical treatment. Each activity can be explained on a chemical basis.

SOLUTIONS

The concentrations of the solutions are not critical. One teaspoon of each chemical in 50–100 mL of water will work well. If the color of the developed ink is not intense enough, make more concentrated solutions.

TEACHING TIPS

SAFETY NOTE: Cobalt chloride is toxic. Avoid touching it.

1. Students can try writing with slightly acidic phenolphthalein solution or copper sulfate solution on paper and then spraying the paper with ammonia in the fume hood.
2. You may prefer to provide ferric chloride solution in a separate bottle with a spray pump. Discarded window cleaner bottles and hair spray bottles work well. Two bottles of ferric chloride solution could be passed around the class.
3. When heated, the acids turn brown and the cobalt chloride turns blue.
4. When chemically treated, potassium thiocyanate turns red.
5. Devise other ways to apply the developing solution (e.g., lightly apply with cotton or dab with moistened paper).
6. Students should avoid any spills and wash their hands after using any of the solutions.
7. In the thiocyanate ion (SCN^-), cyanide is chemically bound to sulfur and as such is not toxic.

8. Any of these solutions can be discarded by flushing down the sink.

ANSWERS TO THE QUESTIONS

1. In the order suggested previously: black or brown, blue, and red.
2. See the Reactions section.
3. See the Reactions section.

96. Identifying the Mystery Metal

An important part of chemistry involves the identification of unknown substances, including metal ions. We can easily identify metal ions because most metal ions undergo characteristic reactions with certain solutions. Such reactions include the production of a gas, a color change, and the formation of a precipitate, each of which is easy to observe. In this activity, we will conduct some tests with a few common metal ions to see what colored precipitates they form with three solutions. We can make a chart of these reactions for easy reference. Then, we can identify a "mystery" metal ion by treating it and comparing the results with our chart.

Materials

1. Piece of plate glass, 6 × 6 in.
2. Ammonium carbonate [$(NH_4)_2CO_3$] solution in a dropper bottle.
3. Ammonium sulfide [$(NH_4)_2S$] solution in a dropper bottle.
4. Potassium iodide (KI) solution in a dropper bottle.
5. Separate solutions containing the following ions: silver ions, aluminum ions, copper ions, lead ions, iron(II) or ferrous ions, and iron(III) or ferric ions.

Procedure

1. Draw around the edges of the glass plate on a piece of white paper to produce a 6 × 6-in. square.
2. Label the paper as follows:

	Ammonium sulfide (S^{2-})	Ammonium carbonate $(CO_3)^{2-}$	Potassium iodide (I^-)
Aluminum (Al^{3+})			
Copper (Cu^{2+})			
Iron(II) (Fe^{2+})			
Iron(III) (Fe^{3+})			
Lead (Pb^{2+})			
Silver (Ag^+)			

3. Place the glass plate over the chart you have just made. Be sure that the glass plate is clean and dry.
4. Observe and record the color of each metal ion solution.
5. Add 2 drops of the solution containing the aluminum ion (Al^{3+}) across from aluminum on the chart and beneath sulfide, beneath carbonate, and beneath iodide.
6. On the aluminum line, beneath the sulfide, add 1 drop of ammonium sulfide solution to the 2 drops already there. What do you observe? Beneath the carbonate, add 1 drop of ammonium carbonate. What do you observe? Beneath the iodide, add 1 drop of potassium iodide. What do you observe? Record all of your observations.

7. Repeat the tests for each metal ion on the chart. Be sure to record your observations each time. There are six sets of these observations, three per set. You will have 18 combinations of solutions to observe. Rinse the glass plate according to your teacher's instructions. Wear gloves.
8. Your teacher will give you a sample of an unknown metal ion in solution. Perform the necessary tests, compare your results with the table you have prepared, and identify the mystery metal ion.

Reactions

Fifteen reactions can be observed in this activity! Let's go down the chart, showing the reaction and the color of any precipitate formed, first considering the reaction with ammonium sulfide. Actually, only the sulfide ion, S^{2-}, reacts; thus, we need not include the ammonium ion in the reaction:

$$2Al^{3+}(aq) + 3S^{2-}(aq) + 6H_2O \longrightarrow 3H_2S(g) + 2Al(OH)_3(s) \quad \text{white}$$

$$Cu^{2+}(aq) + S^{2-}(aq) \longrightarrow CuS(s) \quad \text{black}$$

$$2Fe^{3+}(aq) + 3S^{2-}(aq) \longrightarrow S(s) + 2FeS(s) \quad \text{black}$$

$$Pb^{2+}(aq) + S^{2-}(aq) \longrightarrow PbS(s) \quad \text{black}$$

$$2Ag^{+}(aq) + S^{2-}(aq) \longrightarrow Ag_2S(s) \quad \text{black}$$

Now, let's consider the reactions with ammonium carbonate (actually, only the carbonate ion, CO_3^{2-}, reacts):

$$Al^{3+}(aq) + 3CO_3^{2-}(aq) + 3H_2O \longrightarrow 3HCO_3^{-}(aq) + Al(OH)_3(s) \quad \text{white}$$

$$Cu^{2+}(aq) + CO_3^{2-}(aq) \longrightarrow CuCO_3(s) \quad \text{blue}$$

$$Fe^{2+}(aq) + 2CO_3^{2-}(aq) + 2H_2O \longrightarrow 2HCO_3^{-}(aq) + Fe(OH)_2 \quad \text{green}$$

$$Fe^{3+}(aq) + 3CO_3^{2-} + 3H_2O \longrightarrow 3HCO_3^{-}(aq) + Fe(OH)_3 \quad \text{red}$$

$$Pb^{2+}(aq) + CO_3^{2-}(aq) \longrightarrow PbCO_3(s) \quad \text{white}$$

$$2Ag^{+}(aq) + CO_3^{2-}(aq) \longrightarrow Ag_2CO_3(s) \quad \text{white}$$

Finally, let's look at the reactions with iodide (I^-). Only three of the metal ions form a precipitate with iodide:

$$2Cu^{2+}(aq) + 4I^{-}(aq) \longrightarrow 2CuI(s) + I_2 \quad \text{brown (light)}$$

$$Pb^{2+}(aq) + 2I^{-}(aq) \longrightarrow PbI_2(s) \quad \text{yellow}$$

$$Ag^{+}(aq) + I^{-}(aq) \longrightarrow AgI \quad \text{white}$$

Questions

1. Did you get any precipitates other than those indicated in the Reactions section? Did other students obtain the same results?
2. Why was it not necessary to include the ammonium ion when showing the reaction between ammonium sulfide and the metal ions?
3. Can you draw any general conclusions regarding solubility from your observations?

4. A certain metal ion gave the following results:
   With sulfide, it formed a black precipitate.
   With carbonate, it formed a white precipitate.
   With iodide, it formed a yellow precipitate.
   Based on your observations, can you identify this mystery metal ion?

Notes for the Teacher

BACKGROUND

A great deal of chemistry is packed into this activity! Students are required to organize their experiment and devise a method to record and interpret data obtained through observations. Team work, persistence, and careful observations produce good results. Students also get practice with ionic reactions, solubility, and general qualitative analytical technique.

SOLUTIONS

All solutions should be about 0.2 M. You can approximate this concentration with 2 tsp of each compound in 500 mL of water. (NOTE: You should use distilled water for silver compounds.) Several class sets of solutions in small dropper bottles will save time, prevent contamination of large containers of solutions, and make cleanup and storage easier. You can use any soluble compound containing the following ions: Al^{3+}, Cu^{2+}, Pb^{2+}, Fe^{2+} (ferrous salts), Fe^{3+} (ferric salts), and Ag^{+}.

You will also need 0.2 M solutions of ammonium sulfide [7 g of $(NH_4)_2S$ per 500 mL of solution], ammonium carbonate [10 g of $(NH_4)_2CO_3$ per 500 mL of solution], and potassium iodide (16 g of KI per 500 mL of solution).

TEACHING TIPS

SAFETY NOTE: Silver and lead salts are toxic and must be handled carefully and disposed of according to the procedures in Appendix 4B.

1. Using the "spot plate" technique with small drops of liquid reduces the toxicity problems associated with some of these solutions. However, students should still handle the solutions with care. They can dispose of the solutions by rinsing their plates into a large container.
2. Try putting a master chart on the board for recording student data.
3. Silver sulfide is the black tarnish that forms on silverware. See Activity 58 for a chemical way to remove this tarnish.
4. Only the sulfides of ammonium and the alkali metals (e.g., Li, Na, and K) are soluble.
5. Only the carbonates of ammonium and the alkali metals are soluble.
6. Present unknown solutions, each consisting of any one of the metal ions used in the activity.
7. Small drops work best. Large drops tend to flow into each other.
8. Glass plates must be clean for proper drops to form. Students must be careful when washing and drying the glass plates. Watch out for sharp edges! Round edges and corners with a file before using them or use plastic plates.
9. Many of these reactions are part of a traditional qualitative analysis scheme in which unpleasant odors that characterize these reactions are produced.

ANSWERS TO THE QUESTIONS

1. No other precipitates should be observed. Students must not confuse contaminants such as dust with precipitates. Have them check unexpected results with other students.
2. Because ammonium sulfide, ammonium carbonate, and potassium iodide are all soluble, only the sulfide, carbonate, and iodide ions take part in precipitate formation with the metal ions. The other ions are often called spectator ions because they do not take part in the reaction.
3. Most sulfides and carbonates are insoluble. Only a few of the iodides are insoluble. Most sulfides are black.
4. The mystery metal is lead. See the chart for its characteristic precipitates with sulfide, carbonate, and iodide.

97. Making Fingerprints Visible

When we touch things, the ridges of our finger tips often leave impressions. These impressions are called *fingerprints*. We are not often aware that we leave fingerprints, and sometimes objects must be treated chemically to make these prints visible. In this activity we will make a fingerprint on paper and then treat the paper to make the print visible. You might compare your fingerprint with those of other students. You will find that no two are alike.

Materials

1. Unlined white index cards or white paper.
2. Iodine (I_2) crystals.
3. Erlenmeyer flask, 250 or 125 mL.
4. Burner.
5. Tongs.
6. Scissors.

Procedure

1. Cut narrow strips of paper small enough to be held in the neck of the flask.
2. Press one of your fingers firmly on one end of the paper strip.
3. Place an amount of iodine crystals, about the size of a pencil eraser, in the Erlenmeyer flask. Heat gently with a burner or on a hot plate until the iodine begins to vaporize. Be careful. Do not breathe iodine vapor. Perform this in the fume hood.

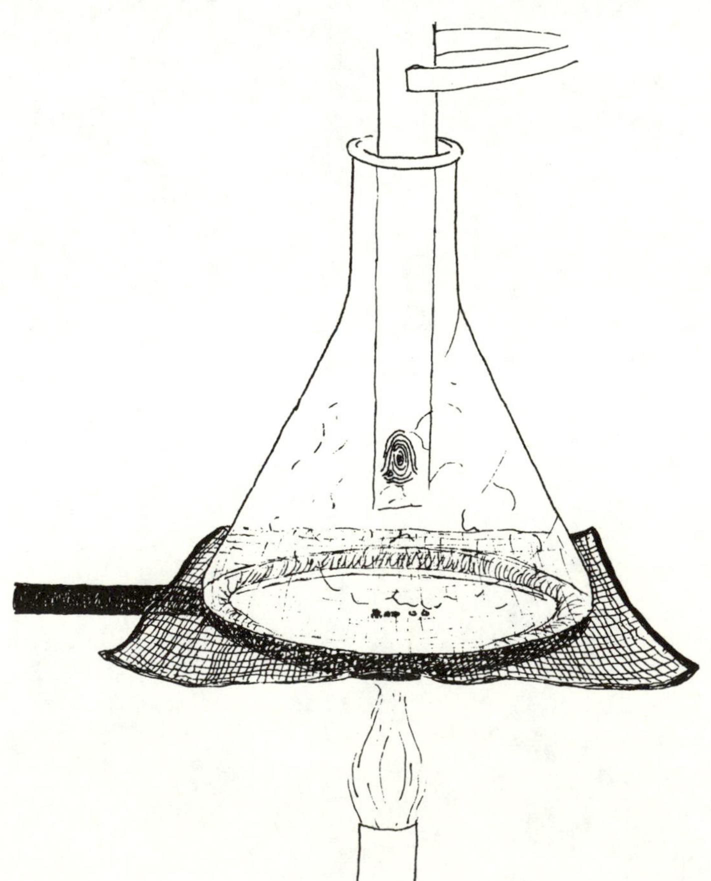

4. Using tongs, dip the paper strips into the flask. What do you observe?
5. Preserve your fingerprint by immediately covering the print with a strip of transparent tape.

Reactions

Fats, fatty acids, and sodium chloride are just a few of the many chemicals that are *exuded* (given off) through the pores of the skin as perspiration. The *exudation* of fats is a natural mechanism for keeping the skin moist and soft. When a fingerprint is made, fat and fatty substances are left behind.

Iodine dissolves in fatty substances and then reacts. Thus, we can produce an outline of the fingerprint by allowing the iodine evaporated from solution to react with the fat. Because only the fat reacts with iodine and turns brown, it will show the ridges of the fingerprint, that is, the area of the finger that made contact with the paper.

Questions

1. Describe what you saw when the fingerprint was exposed to iodine vapor.
2. Try several fingerprints. Are any the same?
3. Try washing your hands with soap and then repeating the activity. What effect did washing have on your ability to detect fingerprints?

Notes for the Teacher

BACKGROUND

This activity involves some organic chemistry, that is, the addition of iodine to a fat molecule. Iodine will react with fatty substances to produce a dark brown compound. We will detect the fingerprint because the ridges of the print are brown.

Materials

Avoid breathing the iodine vapors.

TEACHING TIPS

1. This experiment is a fun exercise, and students enjoy it. You might ask that they all make ONLY thumbprints, and then have a mystery person leave a thumbprint for everyone to identify by comparing it to the class collection.
2. This method is often used now to detect fingerprints. Have students do library research to find other methods.
3. No two people in the world, not even identical twins, have the same fingerprints.
4. The ridges of the finger tips were discovered in 1686 by the Italian anatomy professor Marcello Malpighi.
5. In the mid-1800s, fingerprints were taken by using an imprint of the fingers in printer's ink. This method is still used today.
6. Ridge prints are not limited to the fingers. Babies have their footprints taken shortly after birth. Some dog and cat owners have noseprints made of their pets in case of loss.
7. To get more oil on the finger tip, suggest that students touch their brow before making the fingerprint.

8. Fingerprints left at the scene of a crime are made visible by dusting them with lampblack or a mixture of aluminum powder and chalk dust. The powder sticks to the oily fingerprints.
9. The science of fingerprinting is called *dactyloscopy*. It provides the best means of positive identification known to humans. The only other infallible test is the analysis of blood enzymes. DNA matching is also a current technique. Have students research these new techniques.

ANSWERS TO THE QUESTIONS

1. The outline of the finger tip, provided by the ridges of the print, becomes dark when exposed to iodine vapor.
2. No. No two prints are exactly the same.
3. Answers will vary. Washing may reduce the distinct character of the print; however, the print will probably still be there.

98. Pole-Finding Paper

Place the wire leads from a cell or battery on special paper and watch for the color changes that help you tell which pole is providing electrons. This special paper is heavy filter paper or blotter paper that is soaked with a solution of a salt with phenolphthalein added.

Materials

1. Blotter paper or filter paper.
2. Sodium iodide or potassium iodide.
3. Beaker or cup.
4. Phenolphthalein.
5. A battery or electrical cell (at least 1.5 V) with wires attached to electrodes.

Procedure

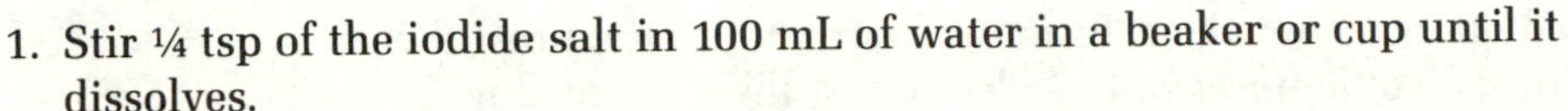

1. Stir ¼ tsp of the iodide salt in 100 mL of water in a beaker or cup until it dissolves.
2. Add 2 drops of phenolphthalein solution to the beaker. The solution should be colorless. If the water is basic, a pink color will form. One drop of vinegar or very dilute acid will eliminate the pink.
3. Soak the special paper in the solution. Make as many pieces of pole-finding paper as you wish with the 100-mL solution.
4. When a piece of paper is damp, place both poles of a cell or battery at least 1 cm apart on the paper. The paper will turn pink at the negative pole where the electrons are entering the solution on the paper. This is the cathode. The other pole will cause a brown color to form.

Reactions

The reactions caused by the electron flow are the reactions for the decomposition of water and oxidation of iodide. At the cathode (negative pole), hydrogen is reduced and OH^- ions are formed. The OH^- causes the phenolphthalein to turn pink:

$$2H_2O + 2e^- \longrightarrow 2OH^- + H_2(g)$$

At the other pole, the anode, iodine is formed from iodide ions:

$$2I^-(aq) \longrightarrow I_2(s) + 2e^-$$

The iodine causes the paper to turn brown where the pole is touching the paper.

Questions

1. How does pole-finding paper work?
2. What reaction makes the pink form on the paper?
3. What reaction makes the brown form on the paper?

Notes for the Teacher

BACKGROUND

If the phenolphthalein on the pole-finding paper turns pink, then the hydroxide ions (OH^-) are present. Electrons are moving into the water and paper at the pole that touches that spot. Then the water molecules break apart to form OH^- and hydrogen gas. Knowing which electrode of a cell is supplying electrons to the circuit is often useful. If you make an electrical cell yourself with metals and solutions, you can find the direction that the electrons are flowing in the circuit. They will be leaving the pole and reacting with the solution where pink occurs and leaving the solution to enter the other pole. Driven by the electrical cell or battery, electrons then flow through the external circuit from the brown to the pink pole.

TEACHING TIPS

1. Pole-finding paper should only be used with small batteries and cells.
2. Students can make predictions about the cells they construct and then test the cells with the paper. Thus, a voltmeter is not needed to test the current flow or the direction of the current.

ANSWERS TO THE QUESTIONS

1. The negative electrode, the cathode, causes the paper to turn pink due to the production of hydroxide ions. The hydroxide ion causes phenolphthalein to become pink.
2. Water molecules being split by electrons.
3. Iodine molecules are formed as electrons leave the iodide ions at the anode. Iodine molecules in water form a brown solution.

Kitchen Chemistry

(These activities are intended to be done outside of the chemistry laboratory.)

99. Chemistry of Cabbage to Sauerkraut

This activity involves an old traditional method for producing sour sauerkraut from sweet cabbage in 2 weeks. The method uses physical changes, osmosis, and diffusion to remove water and sugar from the cabbage. Biochemical changes occur when the sugars are converted to lactic acid by fermenting enzymes in bacteria to produce the acidity that causes the sour, tangy taste. **This activity will take 2 weeks to complete.**

Materials

1. Cabbage.
2. Table salt (NaCl).
3. Universal pH paper or indicator.
4. Clean jar with a lid.

Procedure

PLEASE NOTE: This activity is intended to be done outside of the chemistry laboratory.

1. Shred or finely chop a cabbage.
2. Fill the jar half full with cabbage.
3. Sprinkle about ¼ tsp of table salt over the cabbage.
4. Fill the jar with cabbage.
5. Sprinkle another ¼ tsp of table salt over the cabbage.
6. Slowly add water until the jar is full.
7. Tap the jar to remove air bubbles.
8. Measure the pH with the pH paper or remove a few drops to add to the indicator solution or red cabbage juice indicator.
9. Cap the jar LOOSELY.
10. Allow the cabbage to ferment at room temperature for 2 weeks.
11. Test the pH of your sauerkraut using pH paper or indicator solution from your school laboratory.

Reaction

Making sauerkraut is a kind of fermentation process. The chemical reaction is the enzymatic oxidation of sugars (from the cabbage) to lactic acid:

$$\text{sugar} \xrightarrow{\text{enzymes}} \underset{\text{lactic acid}}{CH_3CHOHCOOH}$$

When you exercise, enzymes in your cells convert glucose to lactic acid by a similar process. The buildup of lactic acid is responsible for the tiredness and soreness of muscles.

Questions

1. Did the pH (acidity) of the cabbage solution change?
2. Why did you remove the air bubbles and keep the lid on the jar?
3. Find out how enzymes work. (Go to the library!)

Notes for the Teacher

BACKGROUND

Lactic acid is a simple organic acid formed by the enzymatic decomposition and oxidation of sugars. This activity is an example of the reaction of bacterial enzymes and of a practical use of this spontaneous process. Enzymes are catalysts produced in cells. The salt concentration outside the cabbage cells causes water and some dissolved sugar in the cells to move out. The bacteria use the sugar as their nutrient and produce lactic acid as one of the byproducts.

TEACHING TIPS

1. Students may bring samples of their sauerkraut juice to school periodically to test pH and share results.
2. The initial pH should be 7. Final acidic pH may be 4 or lower.
3. Be careful when adding salt because too much salt will inhibit the bacteria that form lactic acid.
4. Dill pickles may be prepared from cucumbers by a similar procedure.
5. You may need to add water to the jars during the 2 weeks.
6. Your sauerkraut may be darker than the commercial product, which is "bleached".
7. Some of your students may want to determine how much acid was formed by the bacteria. Titrate the sauerkraut solution with 0.1 M NaOH and use phenolphthalein as the indicator. See Activities 47 and 55.
8. The production of lactic acid by bacteria is a type of fermentation. Other types of fermentation produce alcohols from sugars.

$$\text{sugar} \xrightarrow{\text{enzyme}} \underset{\text{ethyl alcohol}}{CH_3CH_2OH}$$

ANSWERS TO THE QUESTIONS

1. It became more acidic. A pH value lower than 7 indicates acidity.
2. Air inhibits the lactic acid bacteria.
3. Enzymes are specialized catalysts in biochemical reactions.

100. Making Cheese

In this experiment we will make cheese by following the steps used for centuries. First, the protein in milk is made solid, and then it is separated from the liquid part of milk. The solid part is called *curd*, and the liquid is called *whey*. An enzyme is needed to produce the curd. This enzyme is called *rennin* and was originally found in the fourth stomach of suckling calves. It is now made in laboratories.

Materials

1. Prepared milk, ¾ cup (¾ cup of whole milk plus 1 Tbsp of buttermilk).
2. Burner.
3. Rennin (Junket or Rennilase).
4. Stirrer.
5. Pan.
6. Filter paper.
7. Thermometer.
8. Funnel.

Procedure

PLEASE NOTE: This activity is intended to be done outside of the chemistry laboratory.

1. Place the prepared milk in a pan and allow it to sit for 4 h.
2. Warm the milk to 80–90 °F. Be careful not to overheat.
3. While stirring, add one-half of a rennet tablet to the milk. Break the curd that forms into tiny bits.
4. Increase the temperature to 100 °F, but no higher.
5. Remove the container from the heat and allow it to remain undisturbed for 15 min. Observe what happens.
6. Use a coffee filter to filter the contents of the beaker. Your cheese is on the filter.

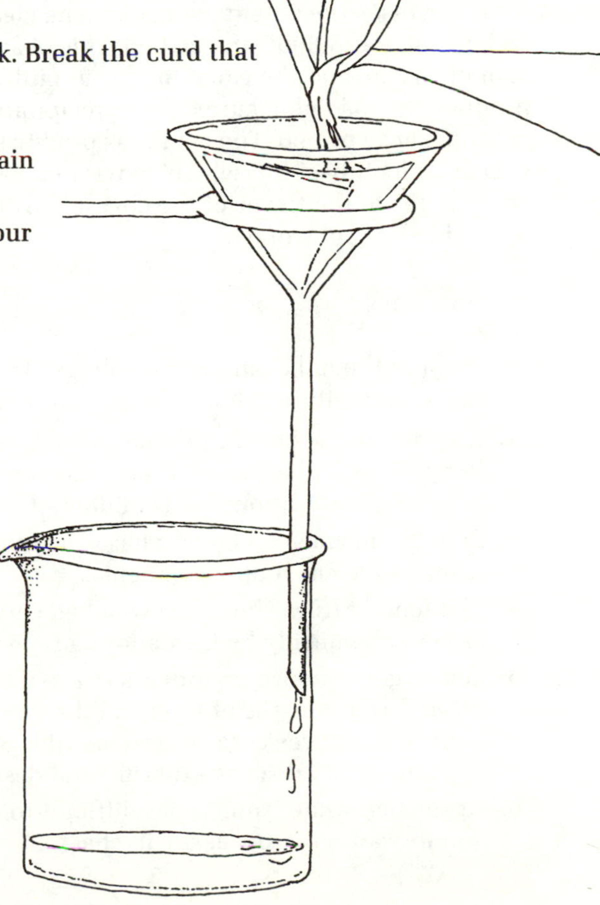

7. Press all of the liquid from the cheese and allow it to dry.

Reaction

1. Souring of milk: The milk sugar called *lactose* is changed to milk acid called *lactic acid* by the action of special bacteria. Lactic acid is a weak acid like the acids found in living tissue.
2. Curdling: Curdling occurs when the enzyme, rennin, is added to milk. The insoluble protein, casein, is formed by this reaction.
3. Aging: Cheese develops flavor during this chemical process as the cheese is stored.

Questions

1. What is in sour milk that makes it sour instead of sweet?
2. Why must the temperature be controlled so carefully?
3. How does your cheese differ from commercially prepared cheese in appearance and taste?

Notes for the Teacher

BACKGROUND

Making cheese is an excellent way to demonstrate enzyme action. The enzyme, rennin, causes the casein molecule to be cleaved into two fractions with molecular weights of about 8000 and 20,000. This change interferes with the previous suspension of casein, and the calcium ions in milk react with the casein fragments to produce the insoluble curds. This precipitation takes place when the milk solution is acidic, near pH 4.6. The acidity is produced by special bacteria, a strain of streptococcus or lactobacillus, which convert lactose to lactic acid. The starter culture also contributes to the flavor of the cheese during *aging*, the chemical process that takes place during cold storage.

TEACHING TIPS

1. Prepare the milk sample by adding 1 Tbsp of buttermilk to each ¾ cup of whole milk. Let it sit for 4 h before using it in this experiment.
2. Emphasize the characteristics of acids as sour and as normal substances in the body. Most biological acids are weak and relatively dilute. Stomach acid, HCl, is an exception. Although it is dilute in the stomach, HCl is a strong acid.
3. Rennin enzyme can be purchased as Junket from grocery stores or as Rennilase from biological supply companies.
4. The temperature should be watched carefully. Enzymes, being proteins, are altered chemically by heat above a range that is optimal.
5. You might encourage students to age their cheeses at different temperatures and for different lengths of time and then make comparisons. Cheese is aged at 2–13 °C for 2 weeks to several months. Such an activity would provide an opportunity to discuss experimental design and variable control.
6. Aged cheeses are sometimes difficult to control because bacteria (both wanted and unwanted) grow easily in cheese.

ANSWERS TO THE QUESTIONS

1. Lactic acid is sour. It is produced from lactose, a sugar, by the action of bacteria.
2. Enzymes will change and no longer work if they are heated too much.
3. The students' cheeses will be soft and white (or gray). They will taste bland. This cheese has not been aged. Commercially, cheese is salted before aging.

101. Mayonnaise: An Edible Emulsion

If you shake an oil and vinegar mixture, you will find it impossible to keep the two liquids mixed together. We call them *immiscible*. The tiny droplets of oil that form will enlarge and eventually become a separate layer. However, the oil droplets can be prevented from coming together if you add a third substance that will stabilize the two liquids in each other when they are shaken together. The third substance is called an *emulsifying agent*, and the mixture of the three substances is called an *emulsion*.

In mayonnaise the three substances are salad oil, vinegar, and the emulsifying agent, namely, egg yolk. You will investigate the activity of the egg yolk in making mayonnaise.

Materials

1. Two egg yolks.
2. Vinegar.
3. Salt.
4. Prepared mustard.
5. Salad oil.
6. Small cup.
7. Small bowl.
8. Electric mixer or beater.
9. Hand lens.

Procedure

PLEASE NOTE: This activity is intended to be done outside of the chemical laboratory.

1. Place vinegar in a small cup to a depth of 1 cm.
2. Add 2 cm of oil to make a total depth of 3 cm in the cup.
3. Stir the contents vigorously in an attempt to mix them.
4. Observe the contents until they are separate again. Which liquid is on the top? Why?
5. Place two egg yolks, ¼ tsp of mustard, ½ tsp of salt, and 1 tsp of vinegar in a clean bowl.
6. Beat the mixture until the egg yolks are sticky.
7. While beating, slowly add 1 cup of salad oil and 2 tsp of vinegar.
8. Continue beating until the emulsion is stable.
9. Use a hand lens to examine the product.

Reaction

The lecithin in the egg yolk acts as the emulsifying agent. The lecithin molecules form a layer around the oil droplets, which prevents the tiny droplets from coming together to form larger drops. Lecithin and other emulsifying molecules have hydrophobic regions that associate with oil, which is also hydrophobic (water-fearing). The other end of the molecule of an emulsifier is usually hydrophilic (water-loving) and associates with water. The emulsifier forms a bridge between water and oil molecules.

Questions

1. When you shook the two liquids in the tube, which one remained on top? Why?
2. How does the appearance of your mayonnaise differ from the commercial product?
3. Describe what you observed through the hand lens.
4. What other products are emulsions? (Go to the library!)

Notes for the Teacher

BACKGROUND

Emulsions are common substances. An emulsion is not a solution; it is a dispersion of one liquid in another and contains an emulsifier for stabilization. Milk is an emulsion of butterfat droplets dispersed in water; the protein, casein, acts as the emulsifying agent to keep the ingredients from separating. Bile contains an emulsifying agent that holds fat droplets dispersed so that they can be digested in the small intestine. Soap is a common emulsifying agent that creates an interface between water and tiny oil droplets, which causes the oil to appear to dissolve in the water. Oily dirt can be washed away.

TEACHING TIPS

1. All ingredients for mayonnaise need to be at room temperature.
2. Students may want to involve family members because a lot of beating and pouring is necessary.
3. This activity is a particularly good one for the strong-armed ones in your class. Suggest that they experiment with the rate dependence of energy input. Remind them that this change is a physical one.
4. This activity works well with another physical change, Activity 5, "Behavior of Gases and the Boiling Egg". If all instruments are nonlaboratory equipment and crackers are available, students can enjoy the products of their labors.
5. Mayonnaise has been made by hand for 300 years. It is named after Port Mahon on Minorca to celebrate a French victory over the British. In French families, making mayonnaise is the Sunday tradition of the father, who takes great pride in the skill.
6. Activity 51 provides students with an opportunity to make cleansing cream, another emulsion.

ANSWERS TO THE QUESTIONS

1. The oil remained on top; it is less dense than the water solution, vinegar, which is about 5% acetic acid.
2. It may be more yellow and more oily in appearance.
3. Small oil droplets can be seen.
4. Milk, butter, paint, cold cream, and glue.

102. Making a Chemical Pie

In this activity we will bake a very unusual apple pie. It will be unusual because it does not contain apples! It will taste like apple pie and look like apple pie because our senses will be tricked into thinking that it is apple pie. We will use chemicals to produce the taste and flavor of apples and cracker pieces to resemble chunks of apples. Because our senses are easily tricked, as scientists we must use sensitive instruments to measure changes that occur around us.

Materials

1. Pastry for two piecrusts (top and bottom).
2. Box of crackers (e.g., Ritz).
3. Sugar (sucrose, $C_{12}H_{22}O_{11}$), 1.5 cups.
4. Butter, 1 tsp.
5. Cinnamon.
6. Cream of tartar (potassium bitartrate, $KHC_4H_4O_6$), 1.5 tsp.
7. Pie pan.
8. Large pot.

Procedure

PLEASE NOTE: This activity is intended to be done outside of the chemistry laboratory.

1. Place 2 cups of water in the pot and heat it until it boils.
2. While the water is heating, mix the sugar and cream of tartar in a bowl.
3. Add the mixture to the boiling water, a little at a time, and stir to dissolve completely.
4. Add 20–25 whole crackers, one at a time, to the boiling solution.
5. Boil for about 3 min, but do not stir.
6. Pour the mixture into a pastry-lined pie pan.
7. Sprinkle a small amount of cinnamon on top of the pie filling.
8. Melt the butter and drip it evenly over the filling.
9. Cover with a pastry top. Stick a knife through the top several times to allow steam to escape.
10. Bake the pie in a preheated oven at 450 °F for about 30 min or until the crust is brown.
11. Cool, and enjoy eating your experiment!

Reaction

The cream of tartar produces a weak acid, which combines with other ingredients to produce the tangy taste of apples. This acid, combined with the pieces of solid cracker, closely resembles the taste and appearance of apple pie.

Questions

1. Find out how sucrose (table sugar) and the starch in flour are related.
2. Why does this taste like an apple pie?
3. What happens to the chemicals you eat?

Notes for the Teacher

BACKGROUND

This activity is a good opportunity to stress the fact that all substances and mixtures are chemicals. We tend to associate the term "chemical" with pure substances that are obtained from supply houses in containers labeled with their purity. In real life, chemicals are not considered as such because they usually are found as complex mixtures. Flour is a mixture of polysaccharides (starch) and proteins. Shortening in the piecrust is a mixture of saturated fats containing mostly carbon, hydrogen, and oxygen atoms. Even the cinnamon is a mixture of carbohydrates and oils. Our taste buds respond to the chemicals in this pie in much the same way as they respond to the chemicals that are "apple" in an apple pie.

TEACHING TIPS

1. This activity is a good excuse to provide a treat for your students. Distribute the procedure sheets. Students can astound their families and friends.
2. This pie is amazingly easy to prepare if you have two ready-made pie crusts: One piecrust is used for the top, and the other is used for the bottom.
3. Before distributing procedure sheets, discuss the mixtures that might be "pie".
4. The students can compile a list of the most obvious chemicals in the pie after consulting some biochemistry texts and the Merck Index.
5. You may want to point out how easily our senses are fooled and thus the importance of making careful measurements with instruments in the laboratory.

ANSWERS TO THE QUESTIONS

1. Sucrose and starch (in flour) are both carbohydrates and can be broken down into simple sugar molecules.
2. The combination of chemicals resembles the flavor of apples.
3. The nutrients are released as separate molecules by the chemical reactions in the digestive system. Then they diffuse into the bloodstream from the small intestine and travel to the cells to be built into large molecules again.

103. Put Fizz in Your Root Beer

Root beer is now made from the bark of the wild cherry tree and other flavors such as clove, mint, and vanilla. If the commercial concentrate is added to water and sugar, something is still missing, namely, the fizz. *Fizz* is carbon dioxide forced into solution by some process. The original root beer is made of bark and roots; it is then fermented so that the yeast slowly produces carbon dioxide in the sealed bottle. However, you will put carbon dioxide fizz into your root beer by immersing solid carbon dioxide into it. Solid carbon dioxide is dry ice. You will discover the properties of a solution of carbon dioxide.

Materials per Team

1. Two heavy sandwich-size plastic bags with a zip seal.
2. Red cabbage juice.
3. Vinegar (acetic acid, 5%), 10 mL.
4. Root beer solution, ½ cup.
5. Two small chunks of dry ice.
6. Clean tongs to handle the dry ice.

Procedure

PLEASE NOTE: This activity is intended to be done outside of the chemistry laboratory.

1. Test the acidic solution.
   A. Place ½ cup of tap water in the plastic bag. Add 2–3 drops of red cabbage juice. Observe the color. (Red cabbage juice is made by boiling chopped red cabbage for a few minutes in enough water to cover it.)
   B. Add a little vinegar to the water and observe the color produced in this acidic solution. You should look for this same color in step D.
   C. Empty the plastic bag and rinse it with water.

   D. Repeat steps A and B, but instead of vinegar, add a small chunk of dry ice. What do you observe? How does the color that is produced compare with that produced in step B?
2. Let's make root beer!
   A. Obtain another clean plastic bag.
   B. Add ½ cup of root beer solution to the plastic bag.
   C. Add a small chunk of dry ice and seal the plastic bag almost completely shut.
   D. Let the reaction proceed until all of the dry ice is gone.
   E. Taste the root beer you have just made. Is it acidic? How can you find out?

Reaction

In commercial root beer the carbon dioxide is forced into the solution under pressure. Gases are more soluble in water if the pressure is increased. Therefore, bottled root beer has more fizz than this type. Carbon dioxide reacts with water to make carbonic acid:

$$\underset{}{CO_2(g) + H_2O(\ell)} \longrightarrow \underset{\text{carbonic acid}}{H_2CO_3(aq)}$$

Questions

1. What would happen if the plastic bag were completely sealed with the dry ice in it?
2. What makes soda pop "go flat" if the lid is left off for a time?
3. Think of some other carbonic acid solutions that are part of your everyday life.
4. Find out about the process of fermentation and how it produces carbon dioxide.

Notes for the Teacher

BACKGROUND

Root beer is fun to make because it serves as a link to the past as well as a chance to produce a beverage that, in modern times, is usually associated with commercial bottling only. This activity is a good time to capitalize on the cross-disciplinary nature of the activity by researching and discussing a little history.

SOLUTIONS

1. To make about 1 gal of root beer, mix 12 mL of root beer concentrate, about 400 g of sugar, and 4 L of water. Root beer concentrate is available in the spice section of supermarkets and grocery stores. Follow the directions on the bottle.
2. You will need 1 lb of dry ice. It can be obtained from ice cream parlors or truck stops.

TEACHING TIPS

1. The dry ice vaporizes (sublimes) readily so that storage is difficult. If you must store it overnight, buy 2 lb, wrap it in newspaper, and store it in a polystyrene cooler. Break it with a blunt instrument such as a hammer and HANDLE IT WITH TONGS AND GLOVES. The temperature is −78 °C.
2. See Activity 47, "Acids and Bases in the Bathroom and Kitchen", for ideas using red cabbage juice indicator.
3. Because the pressure is much lower in the plastic bags than in the commercial process, the root beer produced in this activity will have less dissolved CO_2 and, thus, less fizz than commercial root beer. However, students will get the idea.
4. You might want to snip off a small corner of the top of the plastic bag so that it cannot be sealed all the way and explode from the pressure of the carbon dioxide.
5. You could also produce one batch of root beer by gathering the students around a central transparent container of 1 gal of solution, placing the dry ice in the container, and observing as the beverage forms. Students may be aware of the fizzy feeling in their noses if they inhale the vapors.

6. The vapors are water vapor with dissolved carbon dioxide (carbonic acid).
7. You may use a pressure cooker to create the pressure conditions. Be sure that it has an escape valve! Each student can place a cup with solution and dry ice in the pressure cooker before it is sealed. Allow pressure to build for about 20 min.
8. You might try a batch of yeast-produced root beer as well. It requires 4–6 days and provides a chance to discuss growth rates of organisms and respiration reactions. As the yeast live, they make carbon dioxide just like we do. A recipe is on the root beer concentrate container.

ANSWERS TO THE QUESTIONS

1. It would explode.
2. The carbon dioxide is less soluble at the lower pressure (no lid).
3. Rainwater, blood plasma, mineral springs, and pancake batter.
4. Student research.

104. Team Ice Cream

You can cause cream to freeze quickly if you can lower the temperature around it to well below the freezing point of water, 0 °C. The frozen cream will be smooth if it is in motion as it freezes. Here is the team approach to making ice cream that satisfies these two needs: a cold temperature and continuous motion of the cream. You might say, "Rolling ice cream gathers no large ice crystals."

Materials for Four to Six Students

1. Two coffee cans, 1 and 3 lb.
2. Ice cream recipe, 3 cups: To 3 cups of half and half (half milk and half cream), add ¾ cup of sugar and 2 tsp of vanilla extract. Heat the mixture to dissolve the sugar. Do not boil. Let the mixture cool.
3. Rock salt (NaCl), 2 cups.
4. Crushed ice, 2 qt.
5. Masking tape or duct tape.
6. Mittens.
7. Thermometer (that measures down to −20 °C).

Procedure

PLEASE NOTE: This activity is intended to be done outside of the chemistry laboratory.

1. Pour the cooled ice cream mixture described in Materials 2 into the small can until the can is about three-fourths full. Put on the lid and tape around it.
2. Place the small can into the larger can.
3. Add alternate layers of ice and salt so that the amount of ice is about 4 times the amount of salt.

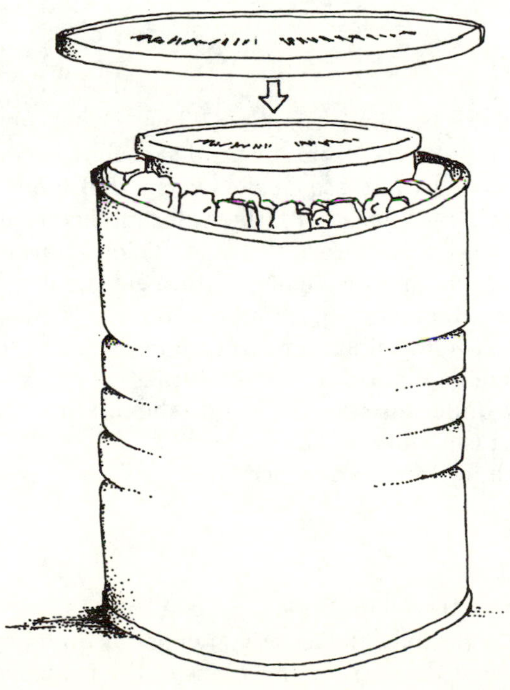

4. Put on the outer lid. Place tape around it.
5. Put on your mittens.
6. Have several students sit in a circle and roll the can back and forth. A towel or foam pad may be wrapped around the can to insulate it.
7. The ice will melt. When very little ice remains, you may add more with a little more salt.
8. After 20–30 min of rolling, open the cans. Enjoy your ice cream!
9. Check the temperature of the water–ice–salt mixture with a thermometer.

Reaction

When the salt dissolves, its particles (ions) interfere with the freezing process, that is, the formation of ice from water. The salt particles become surrounded by water molecules. Even though the temperature is below 0 °C, the ice can melt more easily than the water can freeze because the water molecules are now associated strongly with the salt particles. Energy is needed to melt the ice, and this energy comes from the water. As a result, the water becomes colder. Thus, the salt dissolves in the water, energy goes into melting the ice, and the water–ice–salt mixture gets colder. The mixture should go to about −15 °C. As energy (heat) is removed from the ice cream mixture, it gets very cold and ice cream forms.

Questions

1. Why is the inner can filled only three-fourths full with ice cream mix?
2. How do you explain the rapid increase in the amount of water in the can?
3. What temperature did you find the water–ice–salt mixture to be?
4. How does rolling or churning ice cream make the ice cream smooth rather than full of noticeable ice crystals?

Notes for the Teacher

BACKGROUND

The action of the salt on the water–ice system is complex. When the salt ions [sodium (Na^+) and chloride (Cl^-)] break away from the solid salt, they become surrounded by water molecules. So many of these clusters occur that the water molecules are no longer able to exchange places with the water molecules in the ice, which is the normal case for a water–ice mixture at 0 °C. As the ice keeps melting, it absorbs energy from the liquid water, and the whole mixture gets colder. Melting takes energy. The largest decrease in temperature is found to be −21 °C when 23 g of salt is added to water–ice to make 100 g of mixture. This point is called the *eutectic point.* When making ice cream, the rule of thumb for ice–salt mixtures is a ratio of 4 cups of ice to 1 cup of salt. Because heat leaks into the mixture, the lowest temperature that you might expect is about −15 °C.

Road salt ($CaCl_2$) helps to melt ice by the same principle.

TEACHING TIPS

1. For more "individual" quantities, try smaller cans.
2. The amount of time required to produce the ice cream depends on many variables, including the temperature of the surroundings, size of the ice pieces, amount of salt, and amount of rolling.
3. This activity is nice to do on a festive day, on a spring day outdoors, or as part of a relay event to include rolling the cans down a grassy hill.

4. Duct tape or masking tape around the lids of both cans will prevent them from accidently opening.
5. Principles that are illustrated by this activity include freezing-point depression, ionic solubilities, hydration of ions, and melting as an energy-absorbing (endothermic) process.
6. An extension for investigation could include the study of the lowest temperatures achieved by various proportions of ice and salt.
7. Students might also compare the freezing temperatures observed when salt is dissolved in liquid water. You might try using a dry ice–alcohol slurry in a beaker to find the freezing points of salt solutions of various combinations. The temperature of dry ice is −78 °C.
8. Two standard ice cream recipes will make enough mix for three 1-lb coffee cans. Each can should provide a generous treat for up to six students.

ANSWERS TO THE QUESTIONS

1. As the water in the cream freezes, it takes up 10% more space. Space is needed for movement. Air is also incorporated.
2. Adding salt increases the amount of liquid water.
3. Answers will vary between −10 and −15 °C.
4. As the cream is agitated, the individual crystals forming in the cream are kept small.

105. Making Raisins

The word *raisin* is French and means grape. *Raisins secs* means dried grapes. Most of the raisins produced in the United States are made from the Thompson seedless grape. Four basic processes are used for making raisins. You will investigate one of these processes, namely, making natural raisins. You will compare your natural raisins with commercial raisins produced with sodium hydroxide and sulfur. You will also find the percentage of water in each grape. **This activity will take 4 days.**

Materials

1. Fresh, ripe, seedless grapes.
2. Baking pan.
3. Paper plate.
4. Piece of cheesecloth large enough to cover the paper plate.
5. Warming source (e.g., oven, incubator, heat lamp, or radiator).
6. Balance.

Procedure

PLEASE NOTE: This activity is intended to be done outside of the chemistry laboratory.

1. Artificial drying.
   A. Wash, dry, and determine the mass of 10 grapes as directed by your teacher.
   B. Scald the grapes by dipping them in boiling water until the skin breaks.
   C. Place the grapes in an oven at the lowest setting for 2–4 h until they are soft and leathery.
   D. Check the grapes and shake them gently twice during the drying time.
   E. Cool. Find the mass of the 10 raisins.
2. Raisins in the sun.
   A. Wash and dry 10 of the best grapes.
   B. Find the mass of the grapes.
   C. Place the grapes on a paper plate and cover them with cheesecloth. Staple the cheesecloth to the edge of the plate.
   D. Place the plate in direct sunlight for 4 days.
   E. Determine the mass of the 10 raisins.
   F. Find the average loss of mass per raisin for each process.

Reaction

The making of raisins is called *dehydration*. A similar procedure is followed in chemistry to find the amount of water attached to certain molecules such as blue copper sulfate. The substance is dried and the mass of water is determined by comparing the mass of the substance before and after drying.

Questions

1. What mass of water was lost by the 10 raisins in Procedure 1, "Artificial drying"?
2. What mass of water was lost per raisin in Procedure 1?
3. What mass of water was lost by the 10 raisins in Procedure 2, "Raisins in the sun"?

4. What mass of water was lost per raisin in Procedure 2?
5. For each procedure, determine the percentage of water in a single grape.
6. How do the two methods of making grapes compare?
7. Compare the raisins produced by the two processes that you used to the raisins produced commercially.
8. Devise an experiment to determine the percentage of "water of hydration" in blue copper sulfate.

Notes for the Teacher

BACKGROUND

Four basic processes are used for making raisins:

1. Natural raisins: These are dried in the sun without chemicals. They are dark brown.
2. Golden bleached raisins: Grapes are soaked in 0.5% sodium hydroxide (NaOH) solution to remove the natural wax and to crack the skin. The grapes are then exposed to burning sulfur (SO_2) and dried by artificial heat. These raisins are golden yellow.
3. Sulfur-bleached raisins: Grapes are treated by the same process used to make golden bleached raisins. The only difference is that they are dried in the sun.
4. Lexia raisins: Grapes are dipped in olive oil containing NaOH. Olive oil helps to maintain a tender skin. These grapes are dried in the sun. They are medium to dark brown.

   The water content of many common salts is called the *water of hydration* and is determined by the same method used in this activity. This activity, therefore, serves as an excellent introduction to the idea of percentage of water in any substance and provides practice with the general idea of percentage.

TEACHING TIPS

1. Have students weigh the raisins in the laboratory and take them home in a plastic coffee cup or similar container. They can bring them back for reweighing. Discard raisins after the experiment.
2. By using 10 grapes, students receive practice in thinking in multiples of 10, as in the metric system. In finding the average mass of water lost per grape, see how fast students can divide by 10!
3. You may encounter some practical problems associated with the production of natural raisins. In what parts of the country do you think the process is common? (southern California)
4. Grapes that are dried slowly will lose less water and will have a more leathery appearance than those that are dried in the oven.
5. Dehydrated foods are of interest to hikers, sailors, and those who live in remote areas. Some students may be interested enough to investigate other foods and the efficiency of commercial food dehydrators.

ANSWERS TO THE QUESTIONS

1. This mass is the mass lost per 10 raisins.
2. Divide by 10.
3. See No. 1.
4. See No. 2.

5. Mass of water lost per grape divided by the average mass of one grape.
6. Students should describe the practicality, energy efficiency, and results of the two methods.
7. Have some commercial raisins available.
8. Weigh the blue copper sulfate. Heat in a beaker on a hot plate. Cool. Weigh the copper sulfate again. The copper sulfate is white when it has lost its water of hydration.

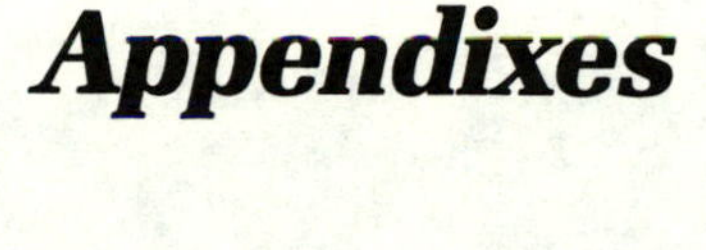
Appendixes

Appendix 1A. Cross Reference of Activities by Chemical Topics

This list suggests appropriate activities for chemical topics most commonly encountered in the chemistry program. The chemical topics are arranged in the same order that they appear in the typical chemistry program.

Chemical Topic	*Activity Number*
Atomic and molecular theory	*1, 8, 9, 11, 17, 25, 26, 38, 43, 44, 49, 88, 89, 90, 91*
Density	*4, 10, 14*
Phase changes	*8, 9, 12, 15, 43, 92, 97*
Gases	*1, 2, 3, 4, 5, 6, 8, 27, 38, 57, 103*
Metals	*14, 15, 22, 23, 24, 27, 35, 37, 40, 42, 58, 63, 91, 96*
Nonmetals	*22, 23, 24, 72, 94, 97*
Crystals	*9, 16, 17, 18, 19, 20, 21*
Colored compounds	*29, 30, 33, 34, 36, 49, 59, 94, 95*
Chemical reactions:	
Gas formation	*4, 7, 27, 38, 39, 61, 82*
Precipitation	*18, 29, 30, 34, 59, 62, 96*
Color changes	*22, 24, 28, 29, 30, 33, 34, 35, 36, 40, 47, 59*
Heat changes	*23, 41, 42*
Reaction rates	*40, 45, 46, 68, 69*
Solutions	*2, 3, 25, 26, 29, 34, 35, 56, 62, 81, 96, 104*
Solubility	*2, 3, 29, 30, 33, 34, 51, 52, 53, 59, 62, 78, 79, 81, 86, 88, 89, 90, 96*
Colloids	*18, 31, 45, 51, 52, 53, 59, 62, 74, 79, 84, 101, 104*
Acids and bases	*4, 6, 7, 28, 33, 47, 55, 56, 57, 66, 67, 71, 77, 82, 98, 99, 103*
Oxidation and reduction	*22, 23, 24, 27, 35, 36, 37, 38, 39, 40, 42, 45, 46, 58, 63, 80*
Chemical energy:	
Electrical	*37, 38, 39, 63, 98*
Heat and light	*23, 36, 41, 42, 104*

Chemical Topic	*Activity Number*
Chemical analysis:	
Qualitative	*7, 34, 48, 57, 70, 72, 75, 76, 77, 78, 79, 81, 84, 87, 88, 89, 90, 91, 92, 95, 96, 97, 98*
Quantitative	*2, 4, 8, 14, 22, 41, 42, 46, 47, 55, 61, 66, 67, 68, 80, 85, 86, 90, 92, 105*
Enzymes	*68, 69, 70, 73, 74, 79, 99, 100*
Polymers	*31, 32, 50, 64, 65, 74, 75, 76, 83*

Appendix 1B. Cross Reference of Activities by Laboratory Process Skills

This list suggests appropriate activities for developing selected laboratory process skills. They are arranged in order of increasing level of reasoning required. Because observing, either through measurement or with the senses, and evaluation of data are included in most of the activities, these two are not part of the list.

Process Skill	*Activity Number*
Classifying	*7, 10, 14, 17, 28, 29, 33, 34, 37, 43, 47, 48, 56, 66, 67, 77, 80, 81, 87, 91, 96*
Recognizing and controlling variables	*2, 5, 7, 9, 11, 18, 22, 26, 28, 33, 40, 41, 42, 43, 45, 46, 47, 48, 55, 56, 62, 63, 66, 67, 68, 69, 70, 78, 79, 80, 81, 85, 86, 87, 89, 90, 100, 105*
Predicting	*7, 8, 9, 18, 24, 26, 29, 30, 33, 34, 37, 42, 46, 47, 48, 56, 63, 66, 67, 68, 73, 74, 79, 81, 86, 87, 96*
Collecting and organizing data in tables	*4, 7, 8, 11, 14, 16, 17, 22, 26, 33, 34, 37, 39, 41, 42, 43, 45, 46, 47, 48, 55, 61, 63, 66, 67, 68, 70, 74, 79, 80, 85, 86, 90, 92, 96, 105*
Analyzing and evaluating data	*4, 6, 11, 14, 16, 17, 22, 23, 24, 26, 33, 34, 37, 39, 41, 42, 43, 45, 46, 47, 48, 55, 61, 63, 66, 67, 68, 70, 74, 77, 79, 80, 85, 86, 90, 96, 105*
Constructing and interpreting graphs	*16, 42, 43, 45, 46, 68, 74, 92*

Appendix 1C. Cross Reference of Activities by Major Topics of *ChemCom* (*Chemistry in the Community*)

ChemCom is a high school science curriculum designed to enhance science literacy by emphasizing the effect of chemistry on society. It was developed by the American Chemical Society and is published by Kendall Hunt Publishing Company, Dubuque, IA 52001.

ChemCom *Section*	*Sequence of Activities in this Book*
UNIT ONE Supplying Our Water Needs	
A. The Quality of Our Water	*94, 62, 12*
B. A Look at Water and Its Contaminants	*49, 11, 104, 29, 30, 95, 88, 89, 90, 92, 48, 77, 78, 34, 96*
C. Investigating the Cause of the Fish Kill	*3, 2, 63, 66, 16, 17, 19, 20, 10, 56, 101*
D. Water Purification and Treatment	*52, 54*
UNIT TWO Conserving Chemical Resources	
A. Use of Resources	*23, 24, 27, 35, 40, 42, 14, 15, 21*
B. Why We Use What We Do	*63*
C. Conservation in Nature and the Community	*22, 61, 65*
D. Metals: Sources and Replacements	*39, 40*
UNIT THREE Petroleum: To Build or To Burn?	
A. Petroleum in Our Lives	*none applicable*
B. Petroleum: What Is It? What Do We Do with It?	*92, 93*
C. Petroleum as a Source of Energy	*52, 41*
D. Making Useful Materials from Petroleum	*31, 32, 51, 54, 82*
UNIT FOUR Understanding Food	
A. Foods: To Build or To Burn?	*102*
B. Food as Energy	*41, 76, 87, 105, 85, 86*
C. Foods: The Builder Molecules	*42, 75, 83, 100, 84*

ChemCom *Section*	*Sequence of Activities in this Book*
D. Substances Present in Foods in Small Amounts	*80, 71, 72, 79, 81*
UNIT FIVE Nuclear Chemistry in Our World	
A. Energy and Atoms	*91*
UNIT SIX Chemistry, Air, and Climate	
A. Life in a Sea of Air	*1, 4, 6, 7*
B. Investigating the Atmosphere	*2, 6, 8, 103, 5, 9, 43, 44, 45, 46*
C. Atmosphere and Climate	*36*
D. Human Impact on the Air We Breathe	*57, 58, 66, 28, 33, 47, 55, 67*
UNIT SEVEN Chemistry and Health	
A. Chemistry Inside Your Body	*25, 26, 68, 69, 70, 73, 74, 99, 50, 100*
B. Chemistry at the Body's Surface	*52, 53, 56, 36, 72*
UNIT EIGHT The Chemical Industry: Promise and Challenge	
A. A New Industry for Riverwood	*52, 53, 59, 64*
B. An Overview of the Chemical Industry	*none applicable*
C. Nitrogen Products and Their Chemistry	*57, 81*
D. Chemical Energy ↔ Electrical Energy	*37, 38, 39*

Appendix 2. Properties and Preparation of Laboratory Acids and Bases

Parameter	Ammonium Hydroxide (NH_4OH)	Acetic Acid ($HC_2H_3O_2$)	Hydrochloric Acid (HCl)	Nitric Acid (HNO_3)	Sulfuric Acid (H_2SO_4)
Dilute this volume (in milliliters) of concentrated reagent to 1 L to make a 1.0 M solution	67.5	57.5	83.0	64.0	56.0
Dilute this volume (in milliliters) of concentrated reagent to 1 L to make a 3.0 M solution	200	172	249	183	168
Dilute this volume (in milliliters) of concentrated reagent to 1 L to make a 6 M solution	405	345	496	382	336
Normality of concentrated reagent	14.8	17.4	12.1	15.7	36.0
Molecular weight	35.05	60.05	36.46	63.02	98.08
Specific gravity of concentrated reagent	0.90	1.05	1.19	1.42	1.84
Approximate percentage in concentrated reagent	57.6	99.5	37.0	69.5	96.0

Note: To make *normal* solutions, use the same amount of reagent shown. However, to make a normal solution of sulfuric acid, use half the amount of reagent indicated. Example: dilute 28.0 mL of concentrated sulfuric acid to make 1 L of 1.0 N sulfuric acid solution.

Safety note: Concentrated acids and bases are corrosive. When you work with 6 M concentrations of acids or bases, wear gloves and safety goggles; work at a safety station equipped with a face shield or free-standing shield. When you dilute acids, always add the acid to water. Caution: Solutions will become hot. Use a fume hood when diluting or using concentrated acids or bases.

Appendix 3. Equipment, Common Materials, and Reagents

This list includes the equipment, common materials, and reagents that are needed for each demonstration. The numbers in italics indicate the number of the demonstration.

Equipment

Simple apparatus and common household materials are used in most of these activities. Common laboratory apparatus include:

Safety goggles
Plastic disposable gloves
Test tubes of various sizes
Test tube racks
Test tube clamps
Beakers, 100 mL to 1 L
Graduated cylinders, 10 mL to 250 mL
Flasks, 125 mL and 250 mL
Funnels
Filter paper
Glass plates
Thermometers
Medicine droppers
Rubber stoppers, solid and with one or two holes
Glass tubing
Rubber tubing
Alcohol or Bunsen burners

Specialized apparatus are useful in the following activities:

Battery *24, 38, 39, 40, 98*
Battery wires and leads *24, 37, 38, 39, 40, 98*
Blender (electric) or egg beater *59, 65*
Bunsen burner or propane torch *91*
Buret *55, 80*
Crucible *15*
Voltmeter or small light bulb *37*
Watch glass *23*
Wire screen, rustproof *64, 65*

Common Materials

Air freshener *9*
Alka Seltzer *4*
Aluminum foil *6, 58*
Ammonia, household *19, 20, 28, 56, 57, 72, 90*
Apples *70, 73, 74, 76*

Baking powder *48, 77*
Baking soda *7, 17, 33, 47, 48, 50, 55, 58, 61*
Banana *87*
Beef, ground *85*
Bleach, laundry *64*
Bluing, laundry *19, 20*
Borax *17, 31, 47, 51, 55, 56*
Boric acid *33, 47*
Bottled soft drinks *2, 81*

Cabbage *99*
Cabbage, red *47, 103*
Chalk, blackboard *89*
Charcoal *59, 94 (activated)*
Chicken bone *71*
Club soda *3, 7*
Coffee *10*
Corn syrup *10, 87*
Crackers *102*
Cream of tartar *102*

Dry ice *60, 83*

Eggs, raw *5, 57, 58, 83, 101*
Eggshells *7*
Epsom salts (magnesium sulfate) *17*

Fats *51*
Flour *65*
Food colors *10, 20, 44, 94*
French fries *86*
Fruit, dried *78*
Fruit juice *70, 80*

Garden fertilizer *57*
Glycerin *53*
Grapes *105*

Hair sample *72*
Hydrogen peroxide *68, 78*

Ink *44, 95*
Iodine, tincture *26, 48, 69, 80, 87*

Laundry starch *48*
Lemon *33, 37, 47, 95*
Limestone *7*
Limewater *2, 6, 7, 61*

Marble chips *7*
Metal samples *14, 15, 18, 22, 23, 35, 37, 42, 46, 58*
Methanol *33, 82, 89*
Milk *50, 59, 79, 84, 100*

Reagents

Appendix 4A. Safe Use of Chemicals

Most chemicals (and all chemicals in this volume) are safe to use **with reasonable care and in the amounts specified.**

Note to the teacher: A few substances fall into special care categories. It is a good idea to have only small amounts of these substances available to students and to keep supplies of larger amounts in a secure place to avoid mishap.

Irritants

Most chemicals can be irritating. Avoid inhaling dusts and fumes of chemicals and avoid spilling chemicals on your skin. Work in the fume hood when you use iodine, ammonia, sodium sulfide, and trichlorotrifluoroethane (TTE). Avoid handling metal chlorides such as those of iron, aluminum, and copper.

Corrosives

Concentrated acids and bases are corrosive. Corrosive substances react with the skin. In the few activities calling for small amounts of 6 M concentration of acids or bases, wear gloves and add the acid or base at a safety station equipped with a face shield or free-standing shield. When you dilute acids, always add the acid to water. Solutions will become hot. Use a fume hood when diluting or using concentrated acids and bases.

Toxic Substances

Some substances should be used in small amounts because they are designated "toxic". Do not put them in your mouth. Such substances include metal salts of barium, lead, cobalt, and silver. See Appendix 4B for proper disposal procedures that prevent these substances from entering the environment.

Note to the teacher: Wear disposable plastic gloves when you mix solutions of these compounds; restrict the amounts available to the students.

Flammable Substances

When you use flammable liquids, extinguish all flames in the laboratory.

Notes to the teacher:

1. The smallest possible amounts of flammable liquids should be available to the students.
2. If your heat source is alcohol burners, take special care. Store replacement fuel (ethyl alcohol) in several bottles that hold 250 mL or less. Store larger quantities in a safety container outside of the laboratory. Fill burners only when they are cool. Fill them at the beginning of class time.
3. Material Safety Data Sheets (MSDS) are provided with orders of chemicals. Save them and keep them close to the laboratory. Each MSDS contains detailed information about the use, toxicity, protective clothing, apparatus, and spill cleanup methods for a specific chemical. As you review these sheets, make them available to interested students in a designated area, not locked up.
4. Learn about American National Standards Institute (ANSI) standards and published recommendations for emergency and safety equipment, such as fume hoods, eyewash fountains, fire extinguishers, and safety goggles. Have the equipment inspected regularly. Safety equipment is available from science supply houses and other commercial sources. Your local and state science supervisor can give you more information.

5. A reference service to help callers find resources and references about health and safety is maintained by the American Chemical Society. The Health and Safety Referral Service can be reached through

   Maureen Matkovich
   American Chemical Society
   1155 16th Street, NW
   Washington, DC 20036
   202-872-4515

Appendix 5 lists other sources of information, including the booklet *Safety in Academic Chemistry Laboratories*, available from the ACS Government Relations and Science Policy Department. Local colleges and universities usually have extensive libraries of books and journals describing accepted practices in use, storage, and disposal of chemicals.

Note: The designations "irritant", "corrosive", and "toxic substance" come from "School Science Laboratories, a Guide to Some Hazardous Substances", available from the U.S. Consumer Products Safety Commission (*see* Appendix 5).

Appendix 4B. Chemical Disposal and Spill Guidelines

Disposal

Metal Compounds

The most environmentally responsible method of disposal of heavy metal compounds is to precipitate them and dispose of them in a managed hazardous waste disposal facility. Your state department of education or local university may have a program through which school laboratory waste materials are collected and managed. Use the following precipitation steps and store the precipitates until you can take them to an approved state or municipal landfill.

Form precipitates of the following metals in a beaker, then filter them and allow them to dry on the filter. Place the dry precipitate and filter paper in a covered jar. Pour the filtrate (liquid) down the drain with running water.

1. **Barium compounds.** Convert barium compounds to barium sulfate compounds. To prepare barium sulfate, add 10% sodium sulfate to a solution of barium salt. This precipitate is very fine. Although soluble barium compounds are toxic, barium sulfate is not considered a hazardous waste.
2. **Lead, cobalt, and silver compounds.** Convert these compounds to metal sulfide compounds. To prepare metal sulfides, add 10% sodium sulfide slowly until precipitation is complete.
3. **Solutions of other metal salts mentioned in this book.** Prepare dilute solutions and pour them slowly down the sink with running water.

Note: Many metal salts not called for in this book require special use and disposal techniques. These include mercury, arsenic, cadmium, chromium, and nickel.

Acids and Bases

Dilute small amounts of acids and bases by adding the acid or base to a large amount of water and pouring the liquid down the drain. Another method is to neutralize acid with dilute aqueous ammonia or baking soda solution. Neutralize bases with dilute hydrochloric acid or vinegar solution.

Organic Compounds

1. **Alcohols, organic acids, and acetone.** Dilute to 10 times their volume with water and then pour them slowly down the drain with running water.
2. **Waste hydrocarbon solvents and esters.** Do not place them in the sink. Consult your school maintenance supervisor for local solvent disposal procedures.

Metals and Nonmetals

Wash and store these substances for reuse. Do not place powdered metals or iodine in the solid waste. Some finely divided metals may react with damp paper to start a fire. Keep used mercury metal in a sealed jar for future reuse.

These disposal guidelines comply with federal regulations. However, your local and state regulations may be different. Obtain information from your state department of environmental quality or from the Environmental Protection Agency (800-231-3075).

Accidental Spills of Irritants, Corrosives, Toxic Substances, or Flammables

Wear gloves and safety goggles. For spills in general, confine the spill, neutralize it, and mop it up.

Confine the spill with cat litter (bentonite). Sand is a less effective alternative.

If an acid spills, neutralize the spill with sodium bicarbonate (baking soda) or sodium or calcium carbonate. Sodium bicarbonate or boric acid can be used to neutralize alkali spills.

Spilled mercury from broken thermometers should be collected and saved in a closed, labeled container. Special mercury spill kits are available from science supply houses. They offer the best assurance of picking up spilled mercury in cracks. Alternatively, dust powdered sulfur onto the spilled droplets and collect all of the material into a jar, capped tightly, to be disposed of as mercury sulfide.

Appendix 5. Useful Resources

American Chemical Society
1155 16th Street, NW
Washington, DC 20036

Education Division

Chem Matters, published four times per year (for high school students). Call (202-872-4590) or write to David Licata to order this magazine or reprints.

Chemunity, published three times per year (free to teachers). This newsletter provides information on ACS activities and other activities in chemical education. Call (202-872-4590) or write to David Licata to order this newsletter.

ChemCom, a chemistry curriculum project developed by ACS for high schools. Call (202-872-4388) or write to Sylvia Ware for information about this publication. To order it, write to Kendall Hunt Publishing Company, Gilam Plaza, Suite 204, 301 East Armour Boulevard, Kansas City, MO 64111.

WonderScience, a chemistry comic book for grade school students in grades 4-6. Call (202-872-6179) or write to Ann Benbow to order this comic book.

Booklets on careers in chemistry are available from Barbara James at 202-872-6168.

Government Relations and Science Policy Department

Safety in Academic Chemistry Laboratories, a general safety manual. Single copies are free; additional copies are $1.00 each. Call or write to Marilyn Charles at 202-452-8917.

Journals Subscriptions and Book Order Department

Chemical and Engineering News, weekly news and chemical applications, special reports.

Chemical Demonstrations: A Sourcebook for Teachers, Volumes 1 and 2.

Journal of Chemical Education. The journal has a section for secondary school chemistry. Reprints are also available. Order from *Journal of Chemical Education*, 20th & Northampton Sts. Easton, PA 18042.

Department of Chemistry
University of Waterloo
Waterloo, Ontario N2L 3G1
Canada

Chem 13 News, a monthly compilation of feature articles and reader contributions in chemistry.

National Science Teachers Association
1742 Connecticut Avenue, NW
Washington, DC 20009

Science and Children, science activities directed to the elementary and middle school student and teacher (grades 1-6).

Science Scope, news and activities for middle school and junior high school teachers (grades 6–9).

The Science Teacher, science and science teaching in the secondary school (grades 9–12).

Journal of College Science Teaching, articles and activities for the college-level science teacher.

U.S. Consumer Products Safety Commission
Office of the Secretary
Washington, DC 20207

"School Science Laboratories, a Guide to Some Hazardous Substances", a booklet, is available free. Call toll-free 800-638-2772 in the continental United States, Hawaii, Alaska, Puerto Rico, and the Virgin Islands or write to the Consumer Products Safety Commission.

Appendix 6. Periodic Chart of the Elements

1	2	3	4	5	6	7	8–10			11	12	13	14	15	16	17	18
IA	IIA	IIIB	IVB	VB	VIB	VIIB	VII			IB	IIB	IIIA	IVA	VA	VIA	VIIIA	NOBLE GASES
1 H 1.0079†																1 H 1.0079	2 He 4.00260
3 Li 6.941†	4 Be 9.01218											5 B 10.81	6 C 12.011	7 N 14.0067	8 O 15.9994†	9 F 18.99840	10 Ne 20.179†
11 Na 22.98977	12 Mg 24.305											13 Al 26.98154	14 Si 28.086†	15 P 30.97376	16 S 32.06	17 Cl 35.452	18 Ar 39.948†
19 K 39.098†	20 Ca 40.08	21 Sc 44.9559	22 Ti 47.90†	23 V 50.9414†	24 Cr 51.996	25 Mn 54.9380	26 Fe 55.847†	27 Co 58.9332	28 Ni 58.70†	29 Cu 63.546†	30 Zn 65.38	31 Ga 69.72	32 Ge 72.59†	33 As 74.9216	34 Se 78.96†	35 Br 79.904	36 Kr 83.80
37 Rb 85.4678†	38 Sr 87.62	39 Y 88.9059	40 Zr 91.22	41 Nb 92.9064	42 Mo 95.94†	43 Tc 98.9064	44 Ru 101.07†	45 Rh 102.9055	46 Pd 106.4	47 Ag 107.868	48 Cd 112.40	49 In 114.82	50 Sn 118.69†	51 Sb 121.75†	52 Te 127.60†	53 I 126.9045	54 Xe 131.30
55 Cs 132.9054	56 Ba 137.34†	57 *La 138.9055†	72 Hf 178.49†	73 Ta 180.9479†	74 W 183.85†	75 Re 186.207	76 Os 190.2	77 Ir 192.22†	78 Pt 195.09†	79 Au 196.9665	80 Hg 200.59†	81 Tl 204.37†	82 Pb 207.2	83 Bi 208.9804	84 Po (210)	85 At (210)	86 Rn (222)
87 Fr (223)	88 Ra 226.0254	89 ‡Ac (227)	104 Unq (261)	105 Unp (262)	106 Unh (262)	107 Uns (262)	108 Uno (265)	109 Une (266)									

*Lanthanide Series

58 Ce 140.12	59 Pr 140.9077	60 Nd 144.24†	61 Pm (147)	62 Sm 150.4	63 Eu 151.96	64 Gd 157.25†	65 Tb 158.9254	66 Dy 162.50†	67 Ho 164.9304	68 Er 167.26†	69 Tm 168.9342	70 Yb 173.04†	71 Lu 174.97

‡Actinide Series

90 Th 232.0381	91 Pa 231.0359	92 U 238.0289	93 Np 237.0482	94 Pu (244)	95 Am (243)	96 Cm (247)	97 Bk (247)	98 Cf (251)	99 Es (254)	100 Fm (257)	101 Md (258)	102 No (255)	103 Lr (256)

Appendix 7. Element Symbols, Atomic Numbers, and Atomic Weights

Name	Symbol	Atomic Number	Atomic Weight	Name	Symbol	Atomic Number	Atomic Weight	Name	Symbol	Atomic Number	Atomic Weight
Actinium	Ac	89	227.03	Helium	He	2	4.00	Radium	Ra	88	226.02
Aluminum	Al	13	26.98	Holmium	Ho	67	164.93	Radon	Rn	86	(222)
Americium	Am	95	(243)	Hydrogen	H	1	1.01	Rhenium	Re	75	186.21
Antimony	Sb	51	121.75	Indium	In	49	114.82	Rhodium	Rh	45	102.90
Argon	Ar	18	39.95	Iodine	I	53	126.90	Rubidium	Rb	37	85.47
Arsenic	As	33	74.92	Iridium	Ir	77	192.22	Ruthenium	Ru	44	101.07
Astatine	At	85	(210)	Iron	Fe	26	55.85	Samarium	Sm	62	150.36
Barium	Ba	56	137.33	Krypton	Kr	36	83.80	Scandium	Sc	21	44.96
Berkelium	Bk	97	(247)	Lanthanum	La	57	138.90	Selenium	Se	34	78.96
Beryllium	Be	4	9.01	Lawrencium	Lr	103	(260)	Silicon	Si	14	28.08
Bismuth	Bi	83	208.98	Lead	Pb	82	207.2	Silver	Ag	47	107.87
Boron	B	5	10.81	Lithium	Li	3	6.94	Sodium	Na	11	22.99
Bromine	Br	35	79.90	Lutetium	Lu	71	174.97	Strontium	Sr	38	87.62
Cadmium	Cd	48	112.41	Magnesium	Mg	12	24.30	Sulfur	S	16	32.06
Calcium	Ca	20	40.08	Manganese	Mn	25	54.94	Tantalum	Ta	73	180.95
Californium	Cf	98	(251)	Mendelevium	Md	101	(258)	Technetium	Tc	43	(98)
Carbon	C	6	12.01	Mercury	Hg	80	200.59	Tellurium	Te	52	127.60
Cerium	Ce	58	140.12	Molybdenum	Mo	42	95.94	Terbium	Tb	65	158.92
Cesium	Cs	55	132.90	Neodymium	Nd	60	144.24	Thallium	Tl	81	204.38
Chlorine	Cl	17	35.45	Neon	Ne	10	20.18	Thorium	Th	90	232.04
Chromium	Cr	24	51.996	Neptunium	Np	93	237.05	Thulium	Tm	69	168.93
Cobalt	Co	27	58.93	Nickel	Ni	28	58.69	Tin	Sn	50	118.69
Copper	Cu	29	63.55	Niobium	Nb	41	92.91	Titanium	Ti	22	47.88
Curium	Cm	96	(247)	Nitrogen	N	7	14.01	Tungsten	W	74	183.85
Dysprosium	Dy	66	162.50	Nobelium	No	102	(259)	Unnilennium	Une	109	(266)
Einsteinium	Es	99	(252)	Osmium	Os	76	190.2	Unnilhexium	Unh	106	(262)
Erbium	Er	68	167.26	Oxygen	O	8	15.999	Unniloctium	Uno	108	(265)
Europium	Eu	63	151.96	Palladium	Pd	46	106.42	Unnilpentium	Unp	105	(262)
Fermium	Fm	100	(257)	Phosphorus	P	15	30.97	Unnilquadium	Unq	104	(261)
Fluorine	F	9	18.998	Platinum	Pt	78	195.08	Unnilseptium	Uns	107	(262)
Francium	Fr	87	(223)	Plutonium	Pu	94	(244)	Uranium	U	92	238.03
Gadolinium	Gd	64	157.25	Polonium	Po	84	(209)	Vanadium	V	23	50.94
Gallium	Ga	31	69.72	Potassium	K	19	39.1	Xenon	Xe	54	131.29
Germanium	Ge	32	72.59	Praseodymium	Pr	59	140.91	Ytterbium	Yb	70	173.04
Gold	Au	79	196.97	Promethium	Pm	61	(145)	Yttrium	Y	39	88.91
Hafnium	Hf	72	178.49	Protactinium	Pa	91	231.04	Zinc	Zn	30	65.38
								Zirconium	Zr	40	91.22

NOTE: Based on carbon-12. Numbers in parentheses are mass numbers of the most stable isotopes of radioactive elements.

Index

Index

E

F

G

H

I

J

K

L

M

Recent ACS Books

Chemical Demonstrations: A Sourcebook for Teachers
By Lee R. Summerlin and James L. Ealy, Jr.
192 pp; spiral bound; ISBN 0-8412-0923-5

Silent Spring Revisited
Edited by Gino J. Marco, Robert M. Hollingworth, and William Durham
214 pp; clothbound; ISBN 0-8412-0980-4

The ACS Style Guide: A Manual for Authors and Editors
Edited by Janet S. Dodd
264 pp; clothbound; ISBN 0-8412-0917-0

Personal Computers for Scientists: A Byte at a Time
By Glenn I. Ouchi
276 pp; clothbound; ISBN 0-8412-1000-4

Writing the Laboratory Notebook
By Howard M. Kanare
146 pp; clothbound; ISBN 0-8412-0906-5

Principles of Environmental Sampling
Edited by Lawrence H. Keith
458 pp; clothbound; ISBN 0-8412-1173-6

Phosphorus Chemistry in Everyday Living, Second Edition
By Arthur D. F. Toy and Edward N. Walsh
362 pp; clothbound; ISBN 0-8412-1002-0

Chemical Reactions on Polymers
Edited by Judith L. Benham and James F. Kinstle
ACS Symposium Series 364; 483 pp; ISBN 0-8412-1448-4

Catalytic Activation of Carbon Dioxide
Edited by William M. Ayers
ACS Symposium Series 363; 212 pp; ISBN 0-8412-1447-6

Pharmacokinetics: Processes and Mathematics
By Peter G. Welling
ACS Monograph 185; 290 pp; ISBN 0-8412-0967-7
